Alexander Yurkin

Symmetric triangle of Pascal and arithmetic parallelepiped

AF138544

Alexander Yurkin

Symmetric triangle of Pascal and arithmetic parallelepiped

On possibility of new evident geometrical interpretation of processes in long pipes

LAP LAMBERT Academic Publishing

Impressum / Imprint

Bibliografische Information der Deutschen Nationalbibliothek: Die Deutsche Nationalbibliothek verzeichnet diese Publikation in der Deutschen Nationalbibliografie; detaillierte bibliografische Daten sind im Internet über http://dnb.d-nb.de abrufbar.

Alle in diesem Buch genannten Marken und Produktnamen unterliegen warenzeichen-, marken- oder patentrechtlichem Schutz bzw. sind Warenzeichen oder eingetragene Warenzeichen der jeweiligen Inhaber. Die Wiedergabe von Marken, Produktnamen, Gebrauchsnamen, Handelsnamen, Warenbezeichnungen u.s.w. in diesem Werk berechtigt auch ohne besondere Kennzeichnung nicht zu der Annahme, dass solche Namen im Sinne der Warenzeichen- und Markenschutzgesetzgebung als frei zu betrachten wären und daher von jedermann benutzt werden dürften.

Bibliographic information published by the Deutsche Nationalbibliothek: The Deutsche Nationalbibliothek lists this publication in the Deutsche Nationalbibliografie; detailed bibliographic data are available in the Internet at http://dnb.d-nb.de.

Any brand names and product names mentioned in this book are subject to trademark, brand or patent protection and are trademarks or registered trademarks of their respective holders. The use of brand names, product names, common names, trade names, product descriptions etc. even without a particular marking in this work is in no way to be construed to mean that such names may be regarded as unrestricted in respect of trademark and brand protection legislation and could thus be used by anyone.

Coverbild / Cover image: www.ingimage.com

Verlag / Publisher:
LAP LAMBERT Academic Publishing
ist ein Imprint der / is a trademark of
OmniScriptum GmbH & Co. KG
Heinrich-Böcking-Str. 6-8, 66121 Saarbrücken, Deutschland / Germany
Email: info@lap-publishing.com

Herstellung: siehe letzte Seite /
Printed at: see last page
ISBN: 978-3-659-38411-0

A. V. Yurkin

Symmetric triangle of Pascal and non-linear arithmetic parallelepiped

On possibility of new evident geometrical interpretation of processes in long pipes

2015

Table of contents

Preface

The offered book is devoted to the evident geometrical description of processes in long tubes.

The book is based on the works published by the author.

1. *A. V. Yurkin.* New mirror for a laser resonator // *Sow. J. Quantum Electron.*, v. 21, p. 447, 1991.

2. *A. V. Yurkin.* Feasibility of reduction laser divergence // Sov. J. Quantum Electron., 1991, v. 21, p. 1096.

3. *A. V. Yurkin.* Geometric features of a laser resonator consisting of many tilted reflecting planes //Sov. J. Quantum Electron., 1992, v. 22, p. 760.

4. *A. V. Yurkin.* Recurrence calculation of laser divergence and refractive analog of a multilobe mirror // Quantum Electron., 1993, v. 23, p. 323.

5. *A. V. Yurkin.* Quasi-resonator a new interpretation of scattering in lasers // *Quantum Electron.*, v. 24, p. 359, 1994.

6. *S. L. Popyrin, I. V Sokolov, A. V. Yurkin.* Three-dimensional geometrical analysis and the characteristics of laser generation in a multilobe mirror cavity // Optics Communications, 1999, v. 164, pp. 297 - 305.

7. *M. B. Mensky, A. V. Yurkin..* The `diffusion' of light and angular distribution in the laser equipped with a multilobe mirror // Procedings of Institute of Systems Analysis of RAS, 2008, v. 32, no. 2, pp. 113 – 121. arXiv:physics/0108037

8. *A. V. Yurkin.* System of rays in lasers and a new feasibility of light coherence control // Optics Communications, 1995, v. 114, p. 393.

9. *A. V. Yurkin.* The ray system in lasers, non-linear arithmetic pyramid and non-linear arithmetic triangles // Proceedings of the Institute of Systems Analysis of RAS, 2008, v. 32, no. 2, pp. 99 – 112 (Russian). arXiv:1302.5214

10. *A. V. Yurkin.* Ray trajectories and the algorithm to calculate the binomial coefficients of a new type // Proceedings of the Institute of Systems Analysis of RAS, 2009, v. 42, no.1, pp. 66 – 77 (Russian). arXiv:1302.4842

11. *A. V. Yurkin.* New view on diffraction discovered by Grimaldi and Gauss

beams // Proceedings of the Institute of Systems Analysis of RAS, 2012,
v. 62, no. 4, pp.28 – 35 (Russian). arXiv:1302.6287

12. *A. V. Yurkin.* New binomial and new view on light theory. About one new universal descriptive geometric model. (Lambert Academic Publishing, 2013). ISBN 978-3-659-38404-2.

It should be noted that the offered book is based, generally on publications [1 - 8] of author which are not presented the book [12].

The offered work was successfully reported and discussed on The 22nd international conference "Mathematics, Computer, Education" in 2015 on January 27 and 28 in the form of poster and oral reports respectively, on the section S1: "Mathematical theories" of http://www.mce.su/. The author is grateful to conferees for attentive and useful discussion of various aspects of work.

The author is grateful to professor E. E. Shnol, to professor V. V. Dikusar, to professor V. G. Mikhalevich to professor A. V. Kaganov, to professor. J. Peters (Canada) and to professor R. Mehta (India) for useful remarks; the author is grateful to professor G. A. Askaryan and academician S. P. Novikov for initial support of the author on this subject more than 20 years ago; the author is also grateful to Mr. N. J. A. Sloane (USA) for his surprising tables.

For convenience of readers, drawings from known books on physics are given in the Appendix and in the work all drawings are original.

1. Introduction

In work [1] a multilobe mirror of the laser resonator for increase of uniformity of laser radiation was offered. In works [2 – 8] theoretical and experimental researches of lasers were presented.

In work [5] the processes taking place in the laser by means of geometrical models of distribution of light were investigated, calculations by means of sequences like Fibonacci's series are carried out.

In work [6] the three-dimensional geometrical model of distribution of a light field in the laser resonator is presented and in work [7] process of "diffusion" of light in the laser is shown.

In work [8] the new evident geometrical model of distribution of light in the laser on the basis of consideration of new binomial distribution is described.

Works [9, 10] are devoted to research of mathematical properties of new binomial distribution. In work [10] the second nonlinear, type of binomial is described, and binomial coefficients, difference from linear Newton binomial is shown.

In work [11] the new view on diffraction of light was offered. Works [9 - 11] can be found in the monograph [12].

In works [4, 5] it was shown that process of distribution of light in lasers can be described in the form of the branching system of links and rays in "binary rays system". Also in work [5] the possibility of the description of such system of rays by means of wavy trajectories ("waves") of length λ_q, where "waves" consist of a set of direct pieces (links) was noted.

In work [8] "the nonlinear arithmetic tree" for the description of distribution of rays of light was offered.

In work [9] the evident model "a nonlinear arithmetic pyramid" for the numerical description of "a nonlinear arithmetic tree" on the basis of consideration of binomial distribution of the 2nd type [10, 13] in paraxial (Gaussian) beams was presented.

In works [9, 11] the assumption was accepted that process of branching of rays happens in open space. Such model can illustrate process of distribution of light or particles in the absence of obstacles or walls.

In works [1 - 12] for the description of processes in lasers, the equivalent "binary rays system" of trajectories of rays (flat waves) inclined under small angles to an axis were offered. It was supposed that rays extend along links of binary rays system and that one ray is proportional to one energy unit.

The present work generalizes above-mentioned previous works of the author.

In the present work we adhere to the evident geometrical approach based on research of properties of systems of straight lines and angles, and also paraxial (Gaussian) approach in calculations.

In the present work the new evident model "a nonlinear arithmetic parallelepiped", for the numerical description of "binary rays system" and for an illustration of process of distribution of the branching system of paraxial (Gaussian) beams and wavy trajectories in long pipes is offered.

The offered new model can be useful to approximate and formal, but evident geometrical interpretation of processes of the movement of particles and waves in long pipes. It is possible to refer to such processes for example the distribution of light in lasers, finding of a particle in infinitely deep potential hole (including new geometrical interpretation of a particle spin), laminar and turbulent flow of liquid in long pipes, etc.

2. Pascal's triangle and arithmetic rectangle

2.1. Pascal's triangle and negative numbers

It is known [14] that binomial coefficients $\binom{n}{p}$ can be calculated by means of the two-dimensional numerical table of an arithmetic triangle of Pascal.

For convenience of the reader we will provide briefly the description of a triangle of Pascal.

Numbers p in this triangle are ordered layer-by-layer in rows, at some distance 2γ from each other, where γ is small distance. The binomial degree n is the ordinal number of a row beginning with the apex of triangle, p is the ordinal number of the numbers in the row and $0 \leq p \leq n$.

Numbers n and p are *positive (natural) integers* and creation of a triangle begins with the top row, with one unit.

We will set the initial conditions for number of a zero row ($n = 0$):

$$\binom{n}{p} = 1 \tag{1}$$

for ($p = 0$), i.e.

$$\binom{0}{0} = 1, \text{ and } \binom{0}{p} = 0 \tag{2}$$

for others p.

We will set also boundary conditions for numbers p of other n - rows:

$$\binom{n}{p} = 0 \tag{3}$$

for $p0$, pn.

Then the rule of consecutive filling with numbers of a triangle of Pascal will be:

$$\binom{n}{p} = \binom{n-1}{p-1} + \binom{n-1}{p}. \tag{4}$$

Or [14]:

$$\binom{n}{p} = \frac{n!}{p!(n-p)!}. \tag{5}$$

We will enter now, for greater *symmetry* of the description of a triangle of Pascal, instead of *positive integers* p, the integers p accepting *positive* or *negative* values, i.e. $p = \cdots, -1, 0, 1, \ldots$, or $p = 0, \pm 1, \pm 2, \ldots$, and $|p| \leq n$. Thus, *integers* of p are ordered in the rows of a *symmetric* triangle of Pascal at distance γ from each other (Fig.1a) i.e. twice more often than *natural* numbers p in a *usual* triangle of Pascal, and the point of $p = 0$ is located on a symmetry axis.

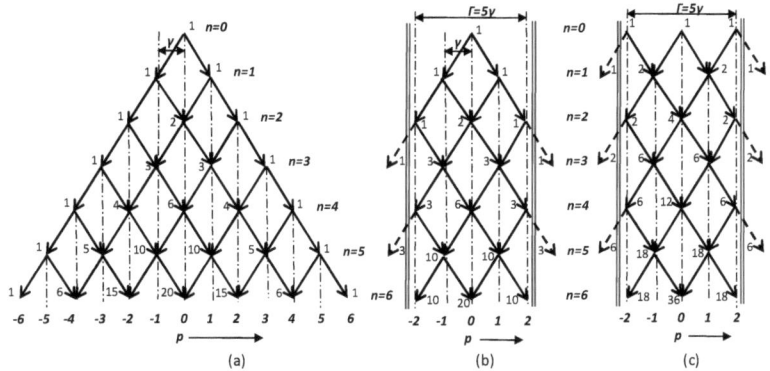

Fig.1. Symmetric triangle of Pascal, ordinal numbers p of integers in a row accept positive and negative values (a); arithmetic rectangle of height of nL, width of Γ; construction begins with one unit (a, b), or with many units (c) of a zero (top) row.

For creation of a symmetric triangle of Pascal, we will set initial conditions for number of a zero row ($n = 0$):

$$\binom{n}{p} = 1, \tag{6}$$

for $p = 0$, i.e.

$$\binom{0}{0} = 1 \text{ и } \binom{0}{p} = 0 \qquad (7)$$

for other p.

We will set also boundary conditions for numbers p of other n - rows:

$$\binom{n}{p} = 0 \qquad (8)$$

for $|p|n$.

Then instead of expression (4) rule of consecutive filling with numbers of a triangle of Pascal will be:

$$\binom{n}{p} = \binom{n-1}{p-1} + \binom{n-1}{p+1}. \qquad (9)$$

And instead of (5) we will have the Newton formula in other form:

$$\binom{n}{p} = \frac{n!}{(\frac{n-p}{2})!(\frac{n+p}{2})!}. \qquad (10)$$

2.2. Arithmetic rectangle

We will consider that our arithmetic rectangle is nL height, and $\Gamma = \gamma m + 1$ width, where L, γ are distances, and m is natural number (Fig.1b, c).

We will set, besides the conditions (8) additional boundary conditions for numbers p of nonzero n - rows:

$$\binom{n}{p} = 0 \qquad (11)$$

for $|p|p_{max}$, where $p_{max} = \Gamma/2$ (Fig.1b).

The rule of consecutive filling with numbers of a triangle of Pascal or arithmetic rectangle will remain the same (9), but process of filling of the two-dimensional numerical table won't go beyond boundary conditions (11) width of Γ (Fig.1b).

We will set additional initial conditions for sequence of numbers of a zero row ($n = 0$) now:

$$\binom{0}{p} = 1, \quad \text{for} \quad |p| \leq p_{max}, \quad \text{and} \quad \binom{0}{p} = 0, \tag{12}$$

for other p.

We will write out sequence of numbers of a zero row in more detail:

$$\binom{0}{0} = 1, \binom{0}{\pm 1} = 1, ..., \binom{0}{|p| \leq p_{max}} = 1, \binom{0}{|p|p_{max}} = 0. \tag{13}$$

In this case the *arithmetic triangle* becomes an *arithmetic rectangle*, and construction begins with the zero ($n = 0$) row consisting of sequence of units (Fig.1c).

The rule of our consecutive filling with numbers of an arithmetic rectangle is (Fig.1c) remains the same (9), taking into account boundary (11) and initial (12, 13) conditions of filling of the two-dimensional numerical table.

It should be noted that at numerical calculations for great values of n the envelope form of distributions of rays in an arithmetic rectangle practically doesn't depend on a type of initial conditions (6, 7) or (12, 13). However, the total number of rays in a case of conditions (12, 13) approximately doubles in comparison with a case (6, 7) (compare Fig.1b with Fig.1c).

2.3. Examples of calculation of an arithmetic rectangle

In Fig.2 the example of calculation of an arithmetic rectangle (Fig.1c) for a case $\Gamma = 7$ according to the rule (9) of consecutive filling with numbers of an arithmetic rectangle taking into account boundary (11) and initial (12 - 14) conditions is given.

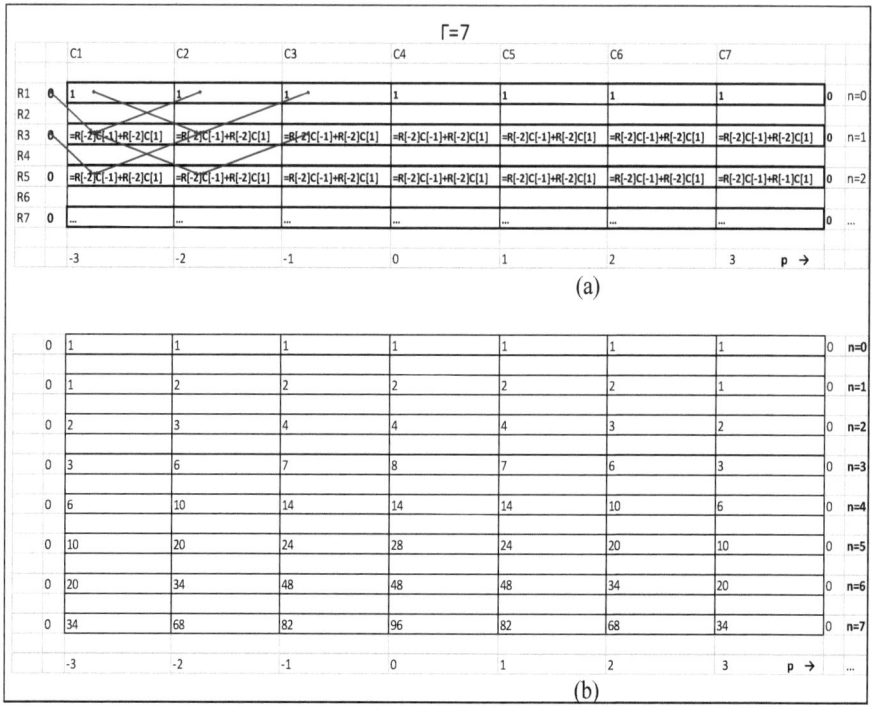

(a)

(b)

Fig.2. Calculation of filling with numbers of an arithmetic rectangle in the Excel program. The table with formulas of calculation (a); by shooters the influencing cells are shown; the number of shooters is approximately doubles in comparison with number of rays in Fig.1c. The same table completed with numerical values (b).

Envelopes schedules of distributions of numbers of rays K for a case $\Gamma = 7$ (Fig.2) are given in Fig.3 at various values $n = 1, 4, 16$ of passes (iterations).

In Fig.3 it is visible that beginning approximately with the 4th pass ($n = 4$) distribution comes nearer to stationary and the envelope form practically doesn't change.

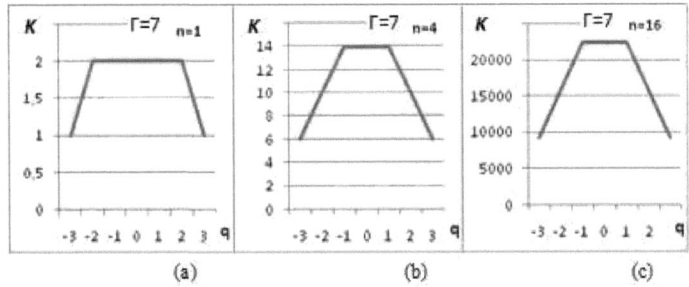

(a)   (b)   (c)

Fig.3. Results of calculation of the arithmetic rectangle given on Fig.2. Envelopes of distributions of numbers of rays K are for $\Gamma = 7$ case. First ($n = 1$) pass (a), fourth ($n = 4$) pass (b) and sixteenth ($n = 16$) pass (c).

Envelopes schedules of distributions of numbers of rays K for a case $\Gamma = 99$ are given in Fig.4 at various values $n = 64, 512, 1019$ of passes (iterations).

In Fig.4 it is visible that beginning approximately with 512 pass ($n = 512$) distribution comes nearer to stationary and the envelope form practically doesn't change and is close to a parabola (Fig.4b, c), and at initial passes the envelope form is close to a rectangle (Fig.4a).

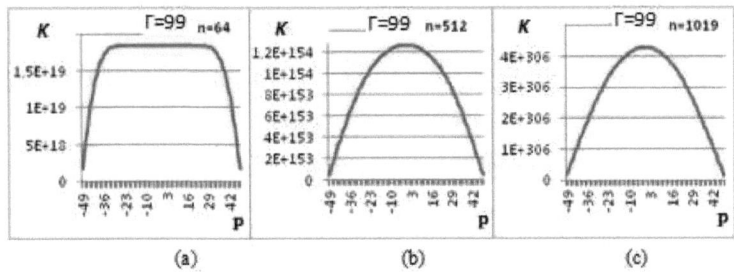

(a)   (b)   (c)

Fig.4. Envelopes of distributions of numbers of rays K for $\Gamma = 99$ case. For 64th ($n = 64$) pass (a), for $n = 512$ (b), $n = 1019$ (c).

3. Nonlinear arithmetic pyramid and nonlinear arithmetic parallelepiped

3.1. Nonlinear arithmetic tree, pyramid and integer rays system

In work [8] the evident geometrical model in the form of a nonlinear arithmetic tree (Fig.5a) was offered. Numbers on this tree are ordered layer-by-layer in rows, at small distance $2k$ from each other. The rays making this tree are inclined on small angles:

$$p = i\gamma \qquad\qquad (14)$$

where $i = 0, \pm1, \pm2 \ldots$ [1, 3, 6, 8, 9]; we will call this group of rays " $i\gamma$ – system" or "integer (ray) system".

In works [9, 12] for the description of this arithmetic tree the three-dimensional arithmetic table in the form of a nonlinear arithmetic pyramid was offered.

For convenience of the reader we will provide briefly [9, 12] description of a nonlinear arithmetic pyramid.

In a nonlinear arithmetic pyramid the numbers are located in the rectangular planes of the different sizes, and the planes are located layer-by-layer one under another, beginning from pyramid top.

Each of n - layers of a pyramid has length of q and width of p. The most top layer ($n = 0$) has zero length of $q = 0$ and zero width of $p = 0$. Numbers n, p, q are natural.

We will designate as

$$\begin{pmatrix} n \\ p \\ q \end{pmatrix} \tag{15}$$

the number located in n a pyramid layer in the rows of p and q.

We will set initial value for number of a zero ($n = 0$) layer, as

$$\begin{pmatrix} n \\ p \\ q \end{pmatrix} = 1 \tag{16}$$

for $p = 0, q = 0$, i.e.

$$\begin{pmatrix} 0 \\ 0 \\ 0 \end{pmatrix} = 1, \text{ and}$$

$$\begin{pmatrix} 0 \\ p \\ q \end{pmatrix} = 0 \tag{17}$$

for other p and q.

We will set also boundary conditions for numbers p and q of other n - layers:

$$\begin{pmatrix} n \\ p \\ q \end{pmatrix} = 0 \tag{18}$$

for $p<0$, pn and $q<0$, $qn(n + 1)/2$.

Then the rule of consecutive filling by numbers of three-dimensional table beginning from the pyramid top will be [9, 12]:

$$\begin{pmatrix} n \\ p \\ q \end{pmatrix} = \begin{pmatrix} n - 1 \\ p - 1 \\ q - p \end{pmatrix} + \begin{pmatrix} n - 1 \\ p \\ q - p \end{pmatrix}. \tag{19}$$

The expression describing the rule of filling with numbers of a nonlinear arithmetic triangle [9, 13] is:

$$\binom{n}{q} = \binom{n - 1}{q - n} + \binom{n - 1}{q}. \tag{20}$$

We will enter now, for greater *symmetry* of the description of a nonlinear arithmetic tree instead of natural numbers p, q the *integers* of p, q accepting positive or negative values; i.e. in the elementary case:

$$p, q = 0, \pm1, \pm2, \ldots, \text{ and } |p| \leq n, |q| \leq n(n+1)/2.$$

Integers of q are ordered in the rows of a *symmetric* nonlinear arithmetic tree at small distance of k from each other (Fig.5a), i.e. twice more often than *natural* numbers q in a *usual* [9, 12] nonlinear arithmetic tree, and the point of $q = 0$ is located on a symmetry axis. *Integers* of p it is displayed in the form of angles, multiple to a small angle γ in the direction (from a vertical) clockwise $(+p)$ or counterclockwise $(-p)$ (Fig.5a); in a *usual* [9, 12] nonlinear arithmetic tree the angles which don't have the direction were postponed.

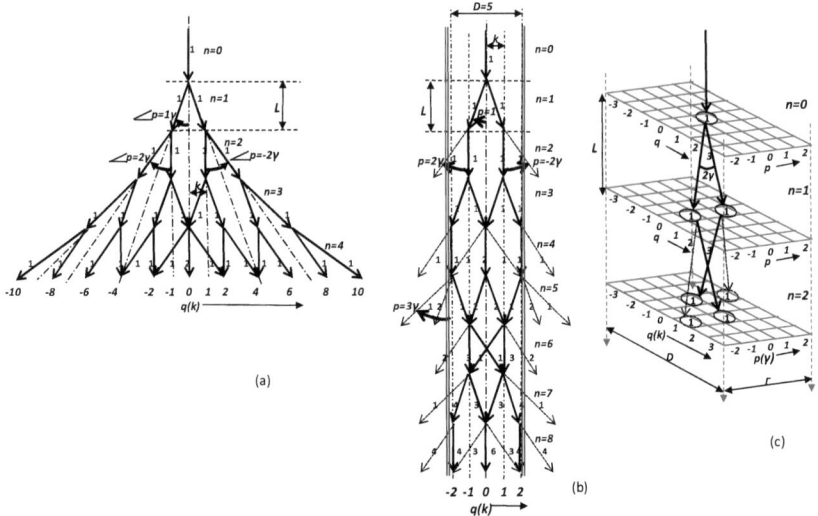

Fig.5. The symmetric nonlinear arithmetic tree, ordinal numbers q of integers in a row accept positive and negative values, numbers p are displayed in the form of the angles having the direction (from a vertical) clockwise $(+p)$ or counterclockwise $(-p)$ (a). The binary ray system, construction begins with one unit (b). The symmetric nonlinear arithmetic pyramid (a nonlinear arithmetic parallelepiped) of nL height, of D length and width of Γ width, construction begins with one unit of a zero (top) row (c).

For creation of *a symmetric nonlinear arithmetic tree* (Fig.5a) and top part of (Fig.5b); and *a symmetric nonlinear arithmetic pyramid* (Fig.5c), we will set initial value for number of a zero ($n = 0$) layer, as

$$\begin{pmatrix} n \\ p \\ q \end{pmatrix} = 1$$

for $p = 0, q = 0$, i.e.

$$\begin{pmatrix} 0 \\ 0 \\ 0 \end{pmatrix} = 1 \text{ и } \begin{pmatrix} 0 \\ p \\ q \end{pmatrix} = 0 \tag{21}$$

for other p and q.

We will set also boundary conditions for numbers p and q for other n - layers:

$$\begin{pmatrix} n \\ p \\ q \end{pmatrix} = 0 \tag{22}$$

for $|p|n$ and $|q|>n(n + 1)/2$.

Then instead of expression (19) rule of consecutive filling with numbers of our tree and the three-dimensional table (Fig.5), since pyramid top, will be:

$$\begin{pmatrix} n \\ p \\ q \end{pmatrix} = \begin{pmatrix} n - 1 \\ p - 1 \\ q + p - 1 \end{pmatrix} + \begin{pmatrix} n - 1 \\ p + 1 \\ q + p + 1 \end{pmatrix}. \tag{23}$$

And instead of expression (20) rule of filling with numbers of a nonlinear arithmetic triangle, will be:

$$\begin{pmatrix} n \\ q \end{pmatrix} = \begin{pmatrix} n - 1 \\ q - n \end{pmatrix} + \begin{pmatrix} n - 1 \\ q + n \end{pmatrix}. \tag{24}$$

3.2. Nonlinear arithmetic parallelepiped

In a nonlinear arithmetic parallelepiped the numbers are located in the rectangular planes of the identical sizes, and the planes are located layer-by-layer one under another beginning from parallelepiped top.

We will consider that our nonlinear arithmetic parallelepiped is of nL height, of $\Gamma = \gamma m + 1$ width and of $D = km' + 1$ length, where L, k are distances, γ is a small angle for our nonlinear arithmetic tree or small distance for our arithmetic pyramid or a parallelepiped, and m, m' are natural numbers (Fig.5c). By means of this model (a nonlinear arithmetic parallelepiped), it appeared, it is possible to describe various types of binary integer rays systems in the long pipes represented, for example as in Figs.5, 6 and 32.

The basic rule of consecutive filling with numbers of a nonlinear arithmetic parallelepiped is the same as at a pyramid, it is described by expression (23).

We will write down, for compactness, the rule (23) of consecutive filling with numbers of a nonlinear arithmetic parallelepiped (Fig.5c) in a form:

$$A = B + C, \tag{25}$$

where $\qquad A = \begin{pmatrix} n \\ p \\ q \end{pmatrix}, B = \begin{pmatrix} n-1 \\ p-1 \\ q+p-1 \end{pmatrix}$ and $C = \begin{pmatrix} n-1 \\ p+1 \\ q+p+1 \end{pmatrix}$.

For creation of various types of nonlinear arithmetic parallelepipeds it is necessary to set various additional boundary and initial conditions.

In Fig.6 different variants of binary ray integer system are presented.

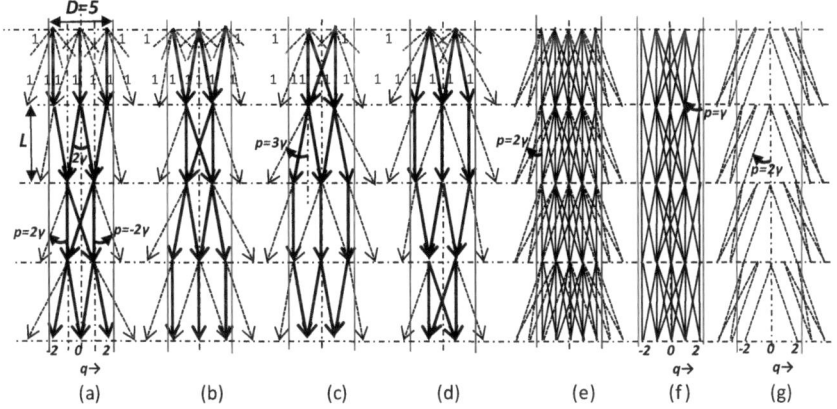

Fig.6. Binary ray $i\gamma$ – system (integer system), height of nL, length of D; rays are inclined on $i\gamma$ angles. Construction begins with many units of a zero (top) row. (a - d) are 4 groups of the rays relating to $i\gamma$ – system, (e) these 4 groups are combined together, (f) the periodic (wavy) trajectories which are a part of $i\gamma$ – the system represented on (e), (g) the no periodic trajectories of $i\gamma$ – system.

We will consider the main types of nonlinear arithmetic parallelepipeds

1. The parallelepiped of type 1.1. Is a case the boundary conditions for numbers q are set and joint calculation of system of periodic (wavy) and no periodic trajectories within length of D of binary rays system and an arithmetic parallelepiped (Figs.5b, c, 6a - e) is made.

In Fig.6a - d are presented 4 groups of rays of $i\gamma$ – system which are described by a nonlinear arithmetic parallelepiped of type 1.1. In Fig.6e all these 4 groups of rays are combined together.

2. The parallelepiped of type 1.2. Is a case the boundary conditions for numbers q are set and joint calculation of system of periodic (wavy) and no periodic trajectories within length of D, and also and no periodic trajectories which are exceeding the bound of length of the D of binary ray system and an arithmetic parallelepiped is made. (Figs.5b, c, 6a - e).

3. The parallelepiped of type 1.3. Is a case the boundary conditions not only for numbers of q, but also for numbers p are set and joint calculation of periodic (wavy) and no periodic systems of trajectories within length of D and width of Γ of binary rays system and an arithmetic parallelepiped is made.

4. The parallelepiped of type 1.4. Is a case the special boundary and initial conditions for numbers q and p are set, and calculation only of periodic (wavy) trajectories within length of D and width Γ of binary rays system and an arithmetic parallelepiped is made.

5. The parallelepiped of type 1.5. Is a case we in common consider two parallelepipeds: type 1.1 and 1.4 and we make only calculation of no periodic systems of trajectories within length of D.

6. We will note that consideration of other various combinations of above-mentioned parallelepipeds is possible.

3.2.1. Parallelepiped of type 1.1

We will set besides the conditions (22) additional boundary conditions for number A in expression (25) for nonzero n - layers:

$$A = 0 \tag{26}$$

for $|q|q_{max}$, where $q_{max} = D/2$ (Fig.5b, c).

Further we will set additional boundary conditions for numbers B and C in expression (25) for nonzero n - layers:

$$B = 0, \text{ and } C = 0 \tag{27}$$

for $|q + p - 1|q_{max}$, and $|q + p + 1|q_{max}$, respectively.

The simplest example of creation of a parallelepiped of type 1.1 for $D = 5$ is given in Fig.5c, in this case the zero layer ($n = 0$) consists only of one unit.

We will set additional initial conditions for sequence of numbers q of a zero layer ($n = 0$) now:

$$\begin{pmatrix} 0 \\ p \\ q \end{pmatrix} = 1$$

for $|q| \leq q_{max}$ and

$$\begin{pmatrix} 0 \\ p \\ q \end{pmatrix} = 0 , \tag{28}$$

for other q.

We will write out sequence of numbers of a zero layer in more detail by analogy with (13):

$$\begin{pmatrix} 0 \\ p \\ 0 \end{pmatrix} = 1, \begin{pmatrix} 0 \\ p \\ \pm 1 \end{pmatrix} = 1, ..., \begin{pmatrix} 0 \\ p \\ |q| \leq q_{max} \end{pmatrix} = 1, \begin{pmatrix} 0 \\ p \\ |q|q_{max} \end{pmatrix} = 0. \tag{29}$$

Not obligatory to set additional boundary conditions for numbers p for a parallelepiped of type 1.1 since numbers p and q are interconnected that follows from expression (23, 25) and boundary conditions (27) for $|q + p - 1|$ and $|q + p + 1|$ numbers.

However for reduction of volume of numerical calculations it is possible to set boundary conditions (similar to conditions (27)) for numbers p considering that for this case as show our numerical calculations:

$$|p| \leq p_{max} \approx 1{,}6\sqrt{D}. \tag{30}$$

Similarly, for reduction of volume of numerical calculations, we will set additional initial conditions for sequence of numbers p of a zero layer ($n = 0$) now:

$$\begin{pmatrix} 0 \\ p \\ q \end{pmatrix} = 1$$

for $|p| \leq p_{max} \approx 1{,}6\sqrt{D}$ and

$$\begin{pmatrix} 0 \\ p \\ q \end{pmatrix} = 0 , \tag{31}$$

for other p.

Thus, the main law of filling with numbers of a nonlinear arithmetic parallelepiped of type 1.1 (Fig.5c) remains, as at a pyramid, former (23, 25), taking into account boundary and initial (26 - 31) conditions of filling of the three-dimensional numerical table.

It should be noted that at numerical calculations for great values of n the envelope form of distributions of rays in a nonlinear arithmetic parallelepiped practically doesn't depend on a type of initial conditions (21) or initial conditions (28 - 31). However, the total number of rays in case of conditions (28 - 31) is approximately quadrupled in comparison with a case (21) (compare Fig.6a - d with Fig.6e).

Examples of calculation of a parallelepiped of type 1.1

In Fig.7 the example of formulas of calculation of a nonlinear arithmetic parallelepiped (Fig.5c) for case of $D = 5$, $\Gamma = 9$ for zero and first passes of system of rays, i.e. $n = 0, 1$ represented in Fig.6e is given. Calculation was made according to the rule (23, 25) of consecutive filling with numbers of an arithmetic parallelepiped taking into account boundary (26, 27) and initial (28 - 31) conditions.

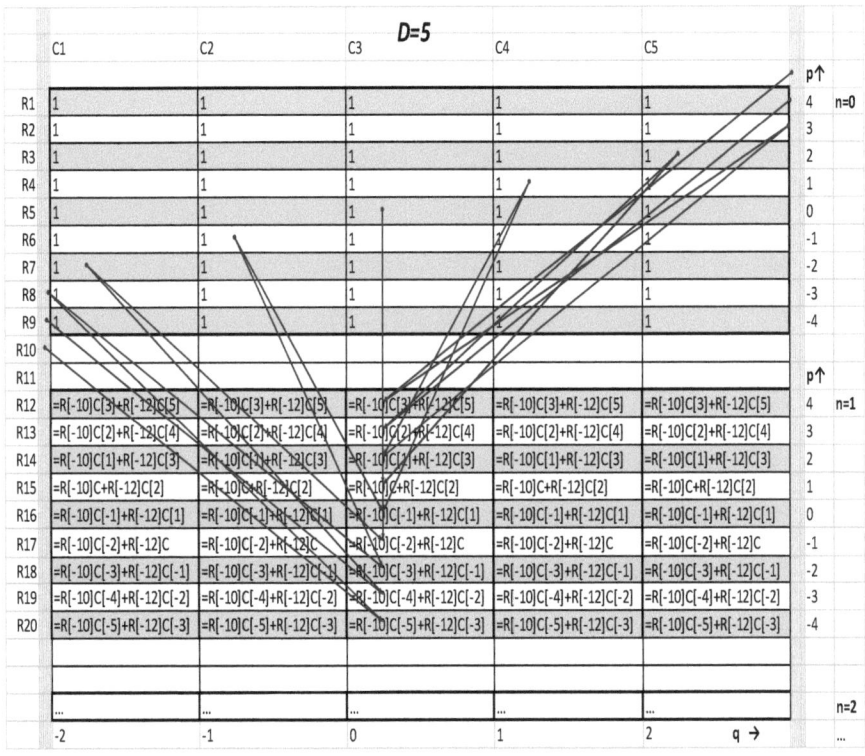

Fig.7. Calculation of filling with numbers of a nonlinear arithmetic parallelepiped of type 1.1 for a case $D = 5$ in the Excel program. The table with the shown formulas is given in drawing, shooters showed the influencing cells.

In Fig.8 the same numerical example for zero, first and 32nd passes of rays, i.e. for $n = 0, 1, 32$ is given. Three rectangles on Fig.8 located from top to down (n = 0, n = 1 and n = 32) are layers of a nonlinear arithmetic parallelepiped (compare Fig.8 with Fig.5c).

D=5; Γ=9

q →	-2	-1	0	1	2		
					n=0		
4	1	1	1	1	1		
3	1	1	1	1	1		
2	1	1	1	1	1		
1	1	1	1	1	1		
↑p 0	1	1	1	1	1		
-1	1	1	1	1	1		
-2	1	1	1	1	1		
-3	1	1	1	1	1		
-4	1	1	1	1	1		
					n=1		
4	1	1					
3	2	1	1				
2	2	2	1	1			
1	2	2	2	1	1		
↑p 0	1	2	2	2	1		
-1	1	1	2	2	2		
-2		1	1	2	2		
-3			1	1	2		
-4				1	1		
...	...	...	...	...	...		
...	...	...	...	...	n=32		
4						$K'(p)$	Σ
3	11144	6528					17672
2	22288	13056	15760	9232			60336
1	26904	22288	22288	15760	15760		103000
↑p 0	22288	22288	31520	22288	22288		120672
-1	15760	15760	22288	22288	26904		103000
-2		9232	15760	13056	22288		60336
-3				6528	11144		17672
-4							
K(q); Σ	98382	89151	107616	89153	98386		

Fig.8. Calculation of filling with numbers of a nonlinear arithmetic parallelepiped of type 1.1 for $D = 5$ in the Excel program. The table is completed with numerical values according to the formulas given on Fig.7.

Envelopes schedules of distributions of numbers of rays (Fig.8) $K(q)$ on a section and $K'(p)$ on a corner of binary rays system for a case of $D = 5$ (Fig.6e) at 32nd pass ($n = 32$) are given in Fig.9. We will note that for this case as show our calculations, the envelope form practically doesn't change approximately after the 15th pass.

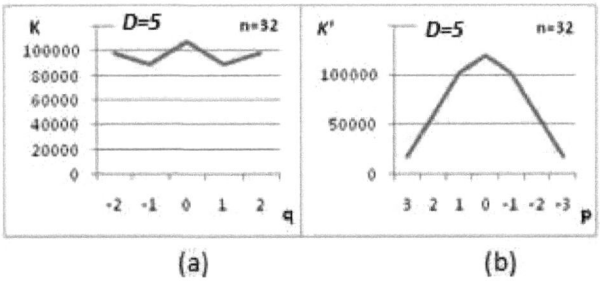

(a)   (b)

Fig.9. Results of filling with numbers of the nonlinear arithmetic parallelepiped shown on Fig.8. The envelopes of distributions of numbers of rays of $K(q)$ on a section (a) and $K'(p)$ on a angle (b) for $D = 5$ case, at the thirty second ($n = 32$) pass.

Envelopes schedules of distributions of numbers of rays of $K(q)$ on a section (Fig.10) and of $K'(p)$ on a corner (Fig.11) of binary rays system for a case $D = 255$ are given in Figs.10, 11 at various values of passes (iterations) of $n = 0, 1, 2, 4, 64, 256$. The envelope form changes for $K(q)$ (Fig.10) from close to a rectangle at initial passes (a – c) to close to a half-ellipse (e, f). The envelope form changes for $K'(p)$ (Fig.11) from close to a rectangle at initial passes (a – c) to close to Gaussian distribution (e, f). Approximately after the 60th passes an envelope form practically doesn't change.

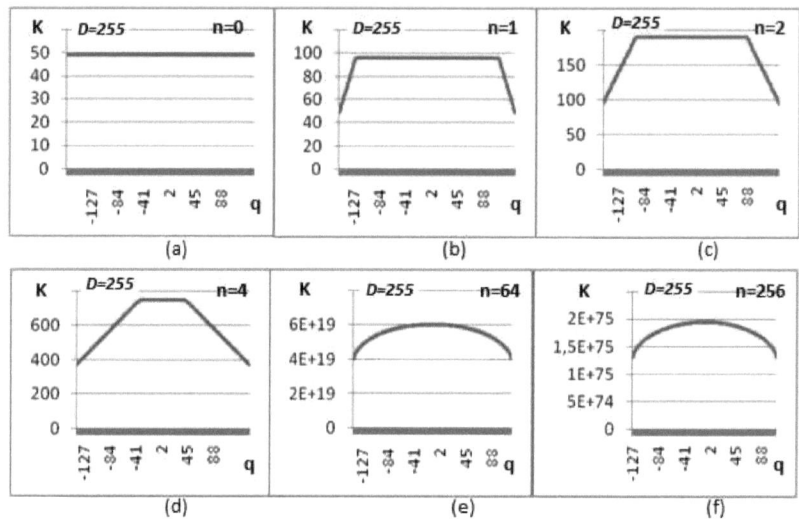

Fig.10. The envelopes of distributions of numbers of rays of $K(q)$ on a section for a case $D = 255$, for pass of $n = 0$ (a), for pass of $n = 1$ (b), for pass of $n = 2$ (c), for pass of $n = 4$ (d), for pass of $n = 64$ (e), for pass of $n = 256$ (f).

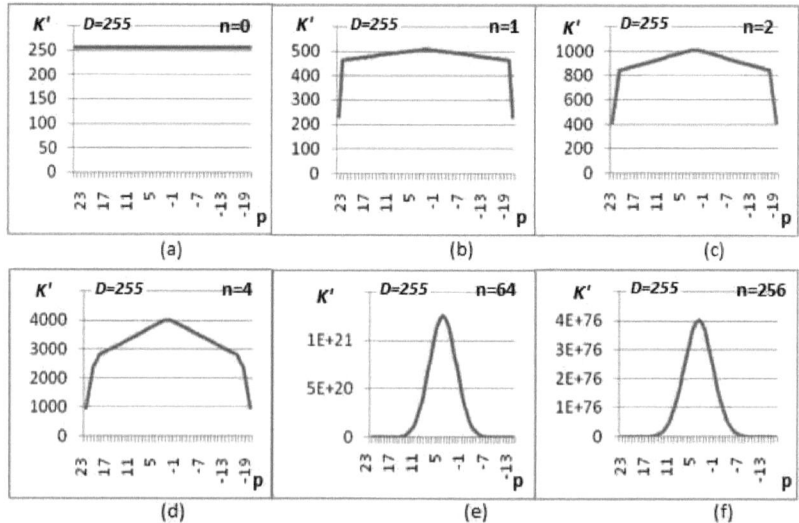

Fig.11. The envelopes of distributions of numbers of rays of $K'(p)$ on an angle for a case $D = 255$, for pass of $n = 0$ (a), for pass of $n = 1$ (b), for pass of $n = 2$ (c), for pass of $n = 4$ (d), for pass of $n = 64$ (e), for pass of $n = 256$ (f).

3.2.2. Parallelepiped of type 1.2

The difference of a parallelepiped of type 1.2 from a parallelepiped of type 1.1 that we don't set additional boundary conditions (26) for number A in expression (25) for nonzero n – layers, and we set only additional boundary conditions (27) for numbers B, and C. It we consider the rays of no periodic trajectories which are going beyond binary rays system.

For reduction of volume of numerical calculations it is possible to set the additional boundary conditions (similar to expression (30)) for numbers p describing the rays of no periodic trajectories which are exceeding the bound of width Γ of a parallelepiped, considering that for this case as show our numerical calculations:

$$|p| \leq p_{max} \approx 1{,}7\sqrt{D}. \tag{30 a}$$

For reduction of volume of numerical calculations it is also possible to set the boundary conditions for numbers q describing the rays of no periodic trajectories which are exceeding the bound of length D of a parallelepiped considering that for this case as show our numerical calculations:

$$|q| \leq q_{max} \approx 1{,}4\sqrt{D} + D/2. \tag{30 b}$$

Examples of calculation of a parallelepiped of type 1.2

In Fig.12 the example of numerical calculation of a nonlinear arithmetic parallelepiped (Fig.5c) for a case of $D = 5$, $\Gamma = 9$ for zero, first and second passes of system of rays, i.e. $n = 0, 1,\ 2$ represented in Fig.6e is given. Rectangles (parallelepiped layers) of number $n = 0$, $n = 1.2$, $n = 2.2\ ...$ coincide with calculation of a parallelepiped of type 1.1 (Fig.8); additional rectangles of $n = 1.1$, $n = 2.1$ belong to a parallelepiped of type 1.2.

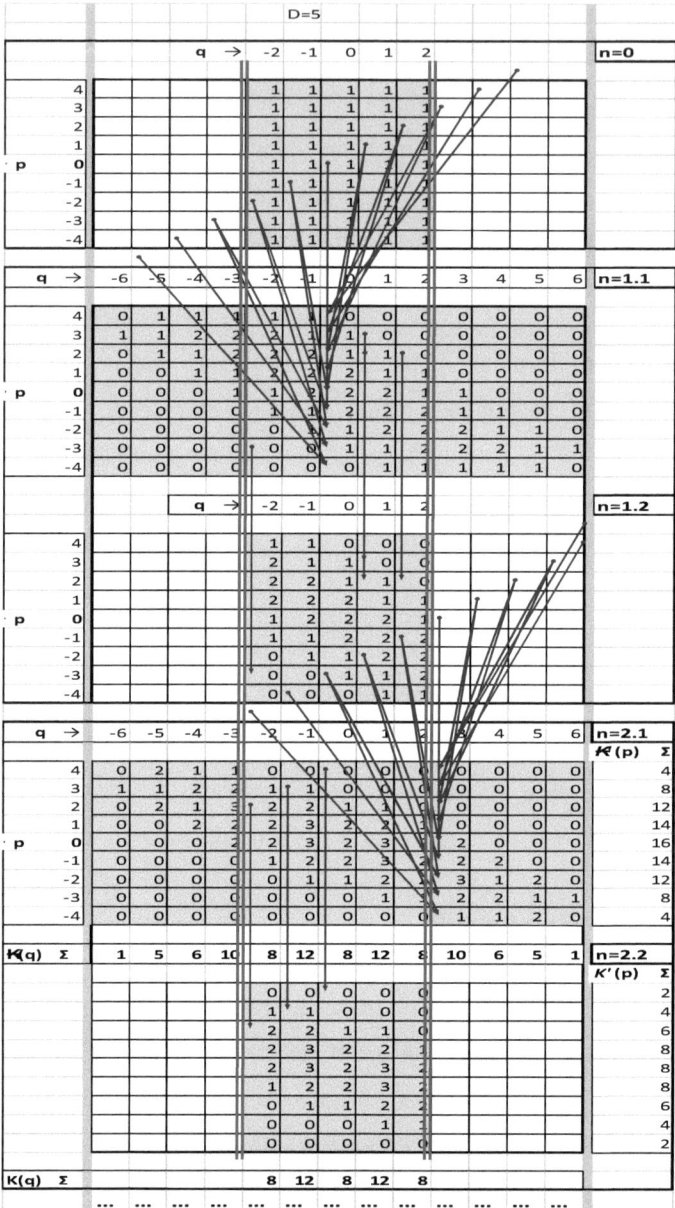

Fig.12. Calculation of filling with numbers of a nonlinear arithmetic parallelepiped of type 1.2 for a case $D = 5$ in the Excel program. The table is completed with numerical values, shooters showed the influencing cells. Number in rectangles of $n = 1.2$, $n = 2.2$ coincide with numbers in the central part of the lines $n = 1.1$, $n = 2.1$.

Envelopes schedules of distributions of number of rays for parallelepipeds of two types are given in Fig.13 in common: (a, b) is for type 1.1 and (c, d) is for type 1.2. Fig.13a, b coincides with Fig.9. $K(q)$, $\mathsf{K}(q)$ are the envelopes on section and $K'(p)$, $\mathsf{K}(p)$ on an angle for parallelepipeds of type 1.1 and type 1.2 respectively. $D = 5$, $n = 32$. We will note that for this case as show our calculations, the form of envelope practically doesn't change approximately after the 15th pass.

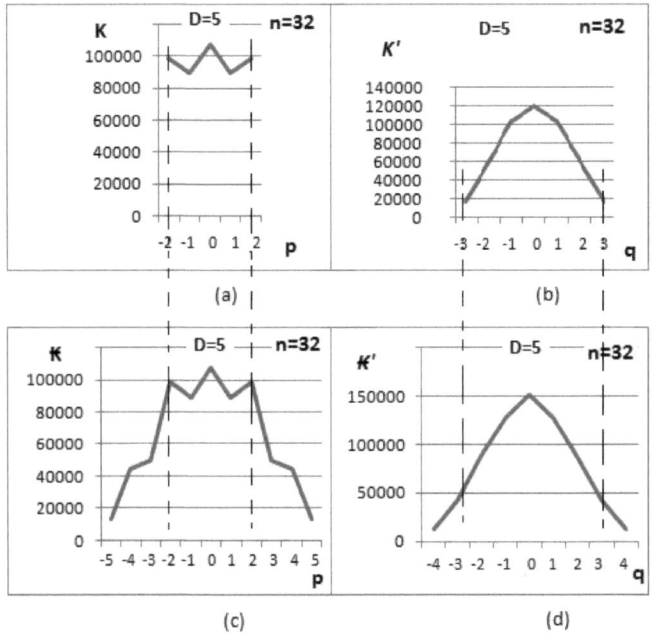

Fig.13. Results of filling with numbers of a nonlinear arithmetic parallelepiped for the case $D = 5$ given on Fig.12. (a, b) are the envelopes of distributions of number of rays for a parallelepiped of type 1.1, (rectangles of $n = 1.2$, $n = 2.2$ in Fig.12) coincide with Fig.9. (c, d) are the envelopes of distributions of number of rays for a parallelepiped of type 1.2 (rectangles of $n = 1.1$, $n = 2.1$ in Fig.12). The envelopes of distributions of number of rays of $K(q)$, $\mathsf{K}(q)$ on a section (a, c) and $K'(p)$, $\mathsf{K}(p)$ on an angle (b, d) for the thirty second ($n = 32$) pass. The dotted line showed border of a pipe aperture of binary ray system for a case $D = 5$, the part of rays of binary rays system (Fig.6) leaves the bounds of an aperture of system (c, d).

Schedules are given in Fig.14 and Fig.15 are similar to that in Fig.13 for $D = 15$, $n = 32$ and $D = 99$, $n = 64$ respectively. We will note that for this case as show our calculations the form of envelope practically doesn't change approximately after the 20th pass for $D = 15$, and after the 30th pass for $D = 99$.

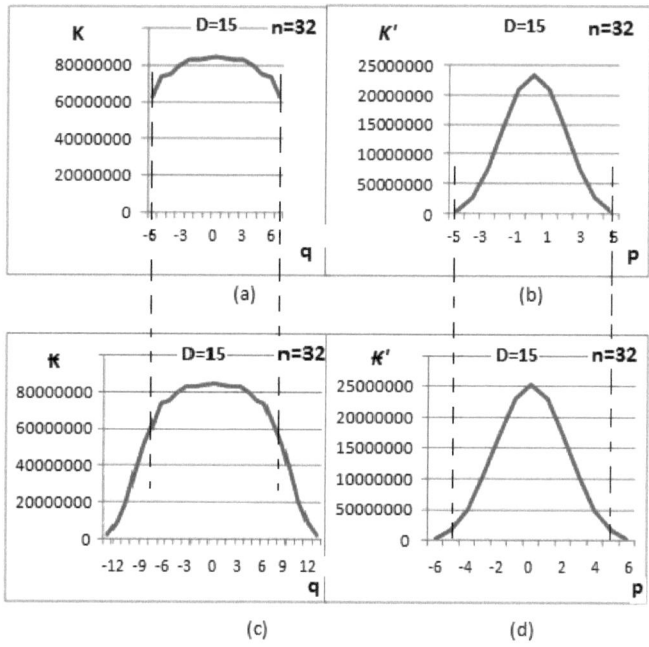

Fig.14. The envelopes of distributions of number of rays of $K(q)$, $K(q)$ on section (a, c) and of $K'(p)$, $K(p)$ on an angle (b, d) for a case of $D = 15$, $n = 32$. The part of rays of binary rays system (Fig.6e, g) leaves the bounds of an aperture (it is marked with a dotted line) of rays system. The forms of envelopes on (b, d), are close to Gaussian distribution, and the part of rays which are going beyond an aperture have an envelope form close to an exponential curve.

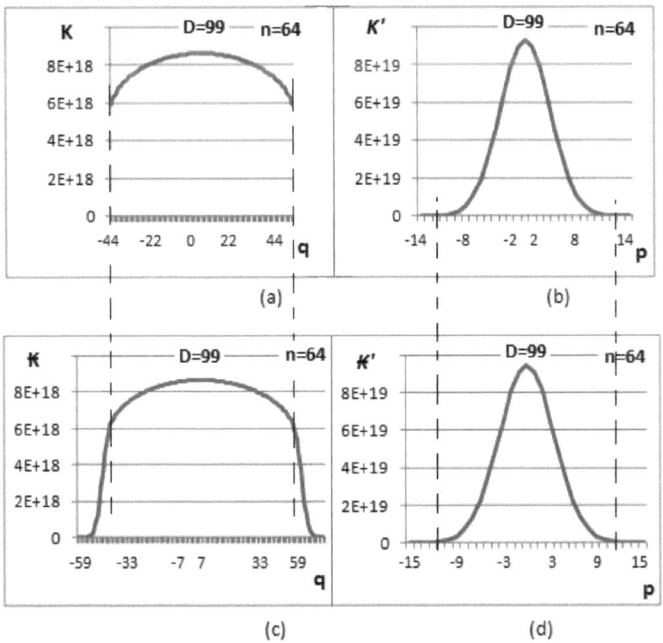

Fig.15. The envelopes of distributions of number of rays of $K(q), Қ(q)$ on section (a, c) and of $K'(p), Қ(p)$ on an angle (b, d) for a case $D = 99$, $n = 64$. The part of rays of binary rays system leaves the bounds of an aperture (it is marked with a dotted line) of ray system. The forms of envelopes on (b, d), are close to Gaussian distribution, and the part of rays which are going beyond an aperture have an envelope form close to an exponential curve.

3.2.3. Parallelepiped of type 1.3

We will set, for a parallelepiped of type 1.1 besides the conditions (26, 27) additional boundary conditions for numbers p of nonzero n - layers:

$$A = 0 \qquad (32)$$

for numbers $|p|p_{max}$ where according to (30) $p_{max} \lesssim 1{,}6\sqrt{D} = \Gamma/2$.

The rule of consecutive filling with numbers of a nonlinear arithmetic parallelepiped will remain the same (23, 25), but process of filling of the three-dimensional numerical table won't go beyond boundary conditions of its length of

D and width of Γ (Fig.5c), taking into account boundary and initial conditions (26, 27, 28 and 32) of filling of the three-dimensional numerical table.

Examples of calculation of a parallelepiped of type 1.3

In Fig.16 (similar to Fig.8) the numerical example of calculation of a nonlinear arithmetic parallelepiped of type 1.3 for a case $D = 5$, $\Gamma = 3$ for zero, first and 32nd passes of rays, i.e. for $n = 0, 1, 32$ is given.

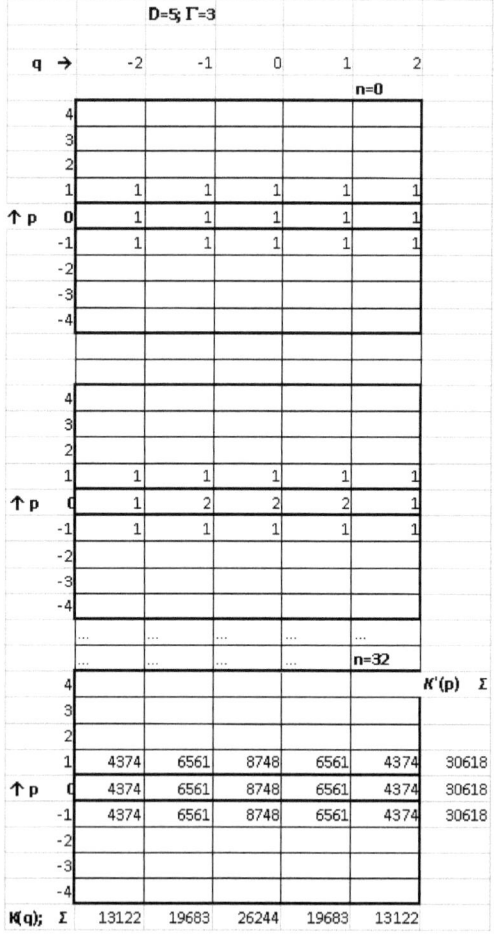

Fig.16. Calculation of filling with numbers of a nonlinear arithmetic parallelepiped of type 1.3 for a case $D = 5$ and $\Gamma = 3$ in the Excel program.

Envelopes schedules of distributions of number of rays in our ray system (Fig.16 parallelepiped of type 1.3) of $K(q)$ on a section and of $K'(p)$ on an angle are given in Fig.17. We will note that for this case ($D = 5$, $\Gamma = 3$) as show our calculations, the form of envelope practically doesn't change approximately after the 15th pass.

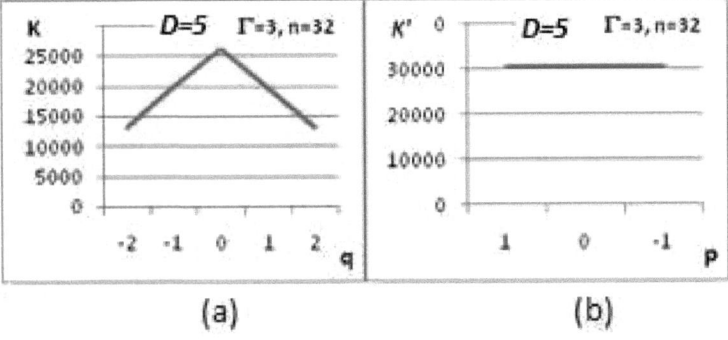

(a)   (b)

Fig.17. Results of calculation of filling with numbers of the nonlinear arithmetic parallelepiped given on Fig.16. The envelopes of distributions of number of rays of $K(q)$ on a section (a) and $K'(p)$ on an angle (b) for a case $D = 5$, $\Gamma = 3$ for the thirty second ($n = 32$) pass.

Envelopes schedules of distributions of number of rays in our ray system of $K(q)$ on a section and of $K'(p)$ on an angle for a case $D = 99$, $\Gamma = 3$ and $n = 64, 512, 1024$ are given in Fig.18. The form of envelope slowly changes for $K(q)$ from close to a rectangle at initial passes to close to a parabola similarly, to the image of envelope (parabola) in Fig.4.

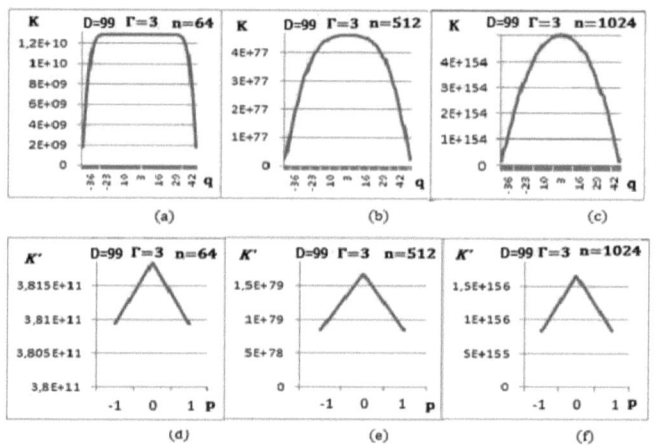

(a)   (b)   (c)

(d)   (e)   (f)

Fig.18. The envelopes of distributions of number of rays of $K(q)$ on a section (a - c) and of $K'(p)$ on an angle (d - f) for a case $D = 99$, $\Gamma = 3$; for $n = 64$ (a, d), for $n = 512$ (b, e) and for $n = 1024$ (c, f).

Envelopes schedules of distributions of number of rays in our ray system of $K(q)$ on a section and of $K'(p)$ on an angle for a case $D = 255$, $\Gamma = 31$ and $n = 256$ are given in Fig.19. The form of envelopes in Fig.19a, b it is close to Fig.10e, f (a part of an ellipse) and Fig.11e, f (a Gaussian curve) respectively. Forms of envelopes in Fig.19c, d are close to parabolas.

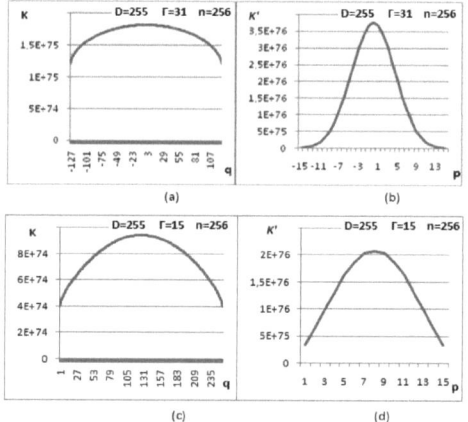

(a)   (b)

(c)   (d)

Fig.19. The envelopes of distributions of number of rays of $K(q)$ on a section (a, c) and of $K'(p)$ on an angle (b, d) for a case $n = 256$, $D = 255$, $\Gamma = 31$ (a, b) and $n = 256$, $D = 255$, $\Gamma = 15$ (c, d).

3.2.4. Parallelepiped of type 1.4

As it was already noted above in item 3.2, a parallelepiped of type 1.4, it is a case at which special boundary and initial conditions for numbers q and p, for carrying out calculation of periodic (wavy) trajectories of binary ray system are set. The example of such system is represented in the form of thick lines of periodic trajectories in Fig.5b and in Fig.6f.

Generally, for construction of such parallelepiped, for each value of numbers q (or groups of such numbers) it is necessary to set boundary and initial values of numbers p.

We will accept that l is some natural numbers of nonzero layers of n depending on numbers (numbers characterizing number of l in a parallelepiped) of n, p, q i.e. $l(n, p, q)$ are the numbers filling a parallelepiped of type 1.4.

We will accept that $l = 1$ for a zero layer ($n = 0$).

We will set boundary and initial conditions for numbers of q_{max} and $-q_{max}$ for all n - layers:

$$\begin{pmatrix} n \\ p \\ q_{max} \end{pmatrix} = l \tag{33}$$

for $p = 0; 1$, and $l = 0$ for other p, and

$$\begin{pmatrix} n \\ p \\ -q_{max} \end{pmatrix} = l \tag{34}$$

for $p = 0; -1$, and $l = 0$ for other p.

For numbers of $q_{max} - 1$, and $q_{max} - 2$:

$$\begin{pmatrix} n \\ p \\ q_{max} - 1 \end{pmatrix} = l, \quad \text{and} \quad \begin{pmatrix} n \\ p \\ q_{max} - 2 \end{pmatrix} = l \tag{35}$$

for $p = -1;\ 0; 1; 2$, and $l = 0$ for other p.

For numbers of $-q_{max} + 1$, and $-q_{max} + 2$:

$$\begin{pmatrix} n \\ p \\ -q_{max} + 1 \end{pmatrix} = l, \quad \text{and} \quad \begin{pmatrix} n \\ p \\ -q_{max} + 2 \end{pmatrix} = l \tag{36}$$

for $p = -2;\ -1;\ 0; 1$, and $l = 0$ for other p.

For numbers of $q_{max} - 3$, $q_{max} - 4$, and $q_{max} - 5$:

$$\begin{pmatrix} n \\ p \\ q_{max} - 3 \end{pmatrix} = l, \quad \begin{pmatrix} n \\ p \\ q_{max} - 4 \end{pmatrix} = l, \quad \text{and} \quad \begin{pmatrix} n \\ p \\ q_{max} - 5 \end{pmatrix} = l \tag{37}$$

for $p = -2; -1;\ 0; 1; 2; 3$, and $l = 0$ for other p.

For numbers $-q_{max} + 3, -q_{max} + 4$, and $-q_{max} + 5$:

$$\begin{pmatrix} n \\ p \\ -q_{max} + 3 \end{pmatrix} = l, \quad \begin{pmatrix} n \\ p \\ -q_{max} + 4 \end{pmatrix} = l, \quad \text{and} \quad \begin{pmatrix} n \\ p \\ -q_{max} + 5 \end{pmatrix} = l \tag{38}$$

for $p = -3; -2 - 1;\ 0; 1; 2$, and $l = 0$ for other p, etc. up to the $|p| \leq p_{max} = \Gamma/2$, where for this case as show our numerical calculations:

$$p_{max} \approx \sqrt{D} \tag{39}$$

Examples of calculation of a parallelepiped of type 1.4

In Fig.20 the numerical example of calculation of a nonlinear arithmetic parallelepiped of type 1.4 for a case $D = 15$, $\Gamma = 9$ for the first three passes of rays, i.e. for $n = 0, 1, 2$ is given.

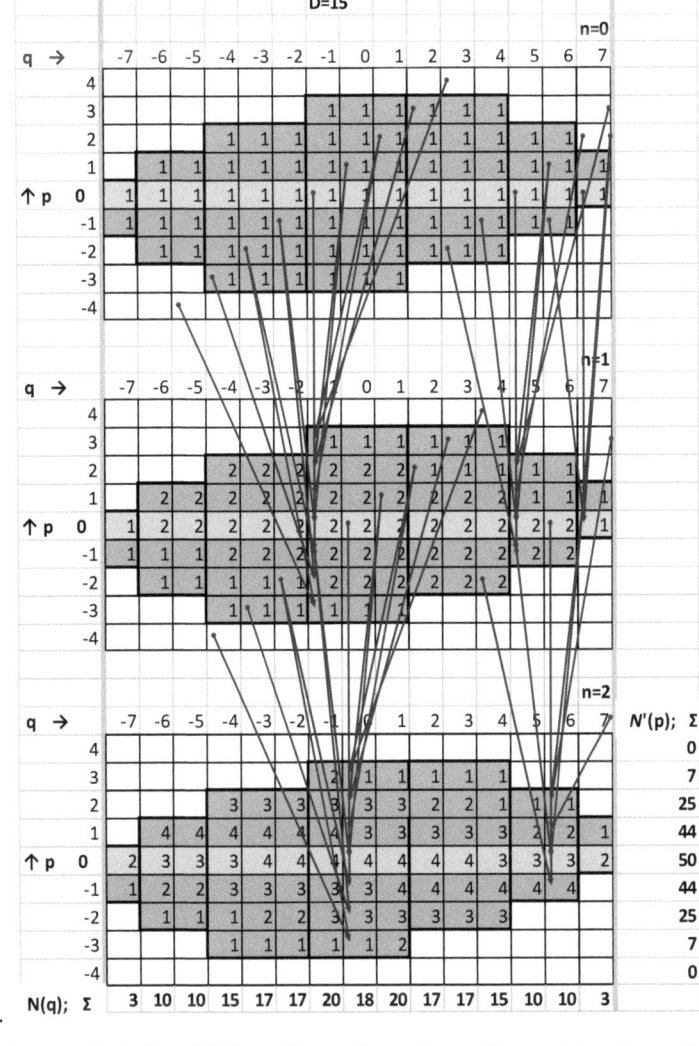

Fig.20. Calculation of filling with numbers of a nonlinear arithmetic parallelepiped of type 1.4 for a case $D = 15$ in the Excel program. Shooters showed the influencing cells.

Envelopes schedules of distributions of number of rays in our ray system (Fig.20 parallelepiped of type 1.4) of N(q) on a section and of $N'(p)$ on an angle are given in Fig.21 for a case $D = 15$, $\Gamma = 9$ and $n = 0, 16, 128$. As show our calculations, the form of envelopes practically doesn't change approximately after the 16th pass.

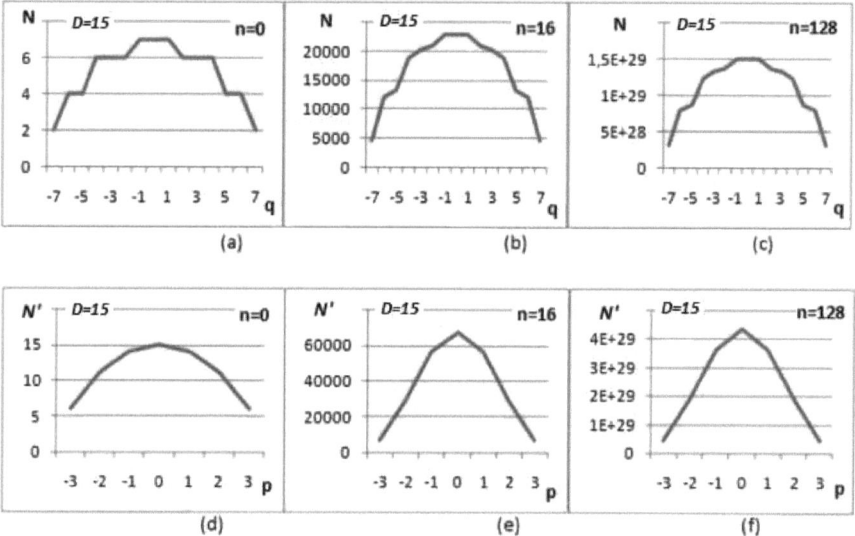

Fig.21. Envelopes of distributions of number of rays of $N(q)$ on a section (a - c) and of $N'(p)$ on an angle (d - f) for a case of $D = 15$ are shown (the calculations are shown in Fig.20).

In Fig.22 (similar to Fig.8) the numerical example of calculation of a nonlinear arithmetic parallelepiped of type 1.4 for a case $D = 5$, $\Gamma = 5$ for zero, first and 32nd passes of rays, i.e. for $n = 0, 1, 32$ is given. In Fig.22 the initial conditions in a form coincide with boundary conditions.

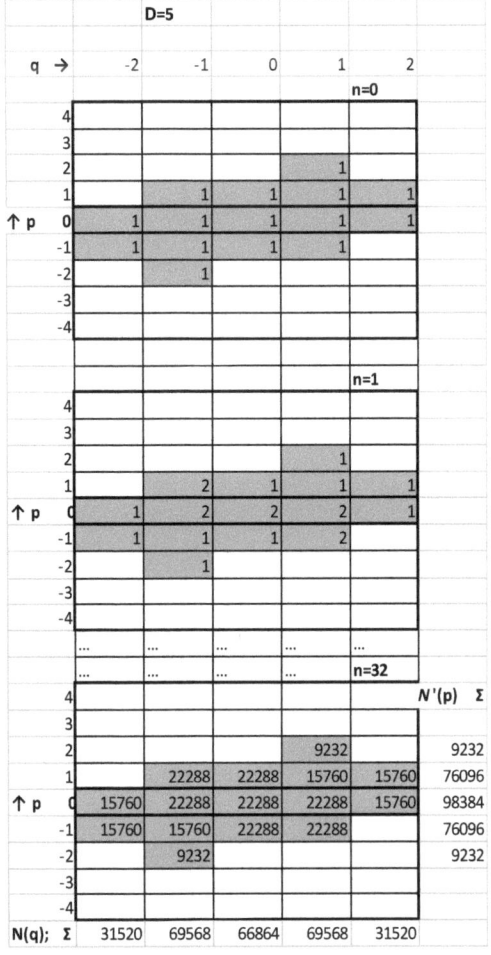

Fig.22. Calculation of filling with numbers of a nonlinear arithmetic parallelepiped of type 1.4 for a case $D = 5$ in the Excel program.

Envelopes schedules of distributions of number of rays in our ray system (Fig.22) of N(q) on a section and of $N'(p)$ on an angle are given in Fig.23 for a case $D = 5$, (Fig.6f) and $n = 32$. As show our calculations, the form of envelopes practically doesn't change approximately after the 15th pass.

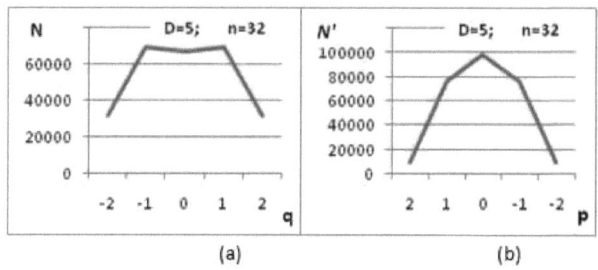

(a)   (b)

Fig.23. The envelopes of distributions of number of rays of $N(q)$ on a section (a) and of $N'(p)$ on an angle (b) for a case of $D = 5, n = 32$ is shown (the calculations are shown in Fig.22).

Envelopes schedules of distributions of number of rays in our ray system (parallelepiped of type 1.4) of $K(q)$ on a section and of $K'(p)$ on an angle are given in Fig.24 for a case $D = 255$ and $n = 0, 64, 256$. The form of the envelope changes for $K(q)$ from close to a parabola of 2nd degree (or parts of an exponential function) at initial passes (a) to close to a parabola of 4th degree (or parts of an exponential function) (b, c). The form of the envelope changes for $K'(p)$ from close to a parabola at initial passes (a) to close to Gaussian distribution (e, f). Approximately after the 60th pass a form of an envelope practically doesn't change.

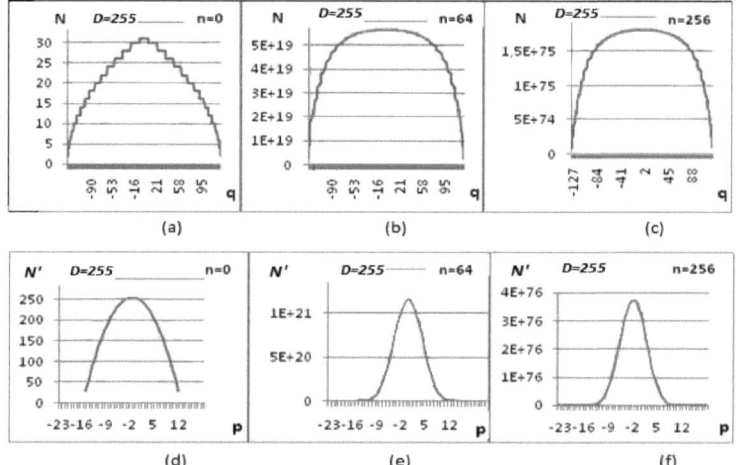

Fig.24 The envelopes of distributions of number of rays of $N(q)$ on a section (a - c) and of $N'(p)$ on an angle (d - f) for a case of $D = 255$.

3.2.5. Parallelepiped of type 1.5

As it was already noted above in item 3.2, a parallelepiped of type 1.5 it is a case at which are in common considered of parallelepipeds of type 1.1 and of type 1.4 for carrying out calculation of no periodic systems of trajectories in binary rays system within length of D (Fig.6g). The parallelepiped of type 1.1 describes the general distribution of K periodic and no periodic trajectories (Fig.6e) Parallelepiped of type 1.4 describes distribution of N only of periodic (wavy) trajectories within length of D (Fig.6f). The parallelepiped of type 1.5 describes distribution only of the M no periodic trajectories within length of D. Therefore, distribution of no periodic trajectories of M can be received from simple expression:

$$M = K - N. \tag{40}$$

Examples of calculation of a parallelepiped of type 1.5

In Fig.25 the numerical example of calculation of a nonlinear arithmetic parallelepiped of type 1.5 for a case $D = 5$ for zero, first and 32nd passes of rays i.e. for $n = 0, 1, 32$ is given.

Fig.25a coincides with Fig.22. The example presented in Fig.25b differs from an example in Fig.8 only on initial conditions. In Fig.25a, b initial conditions in a form coincide with boundary conditions. Results of calculations (Fig.8 and Fig.25b) are close, especially for great values of D and n.

The difference between distribution of numbers N in Fig.25a and K in Fig.25b is shown as M in Fig.25c.

42

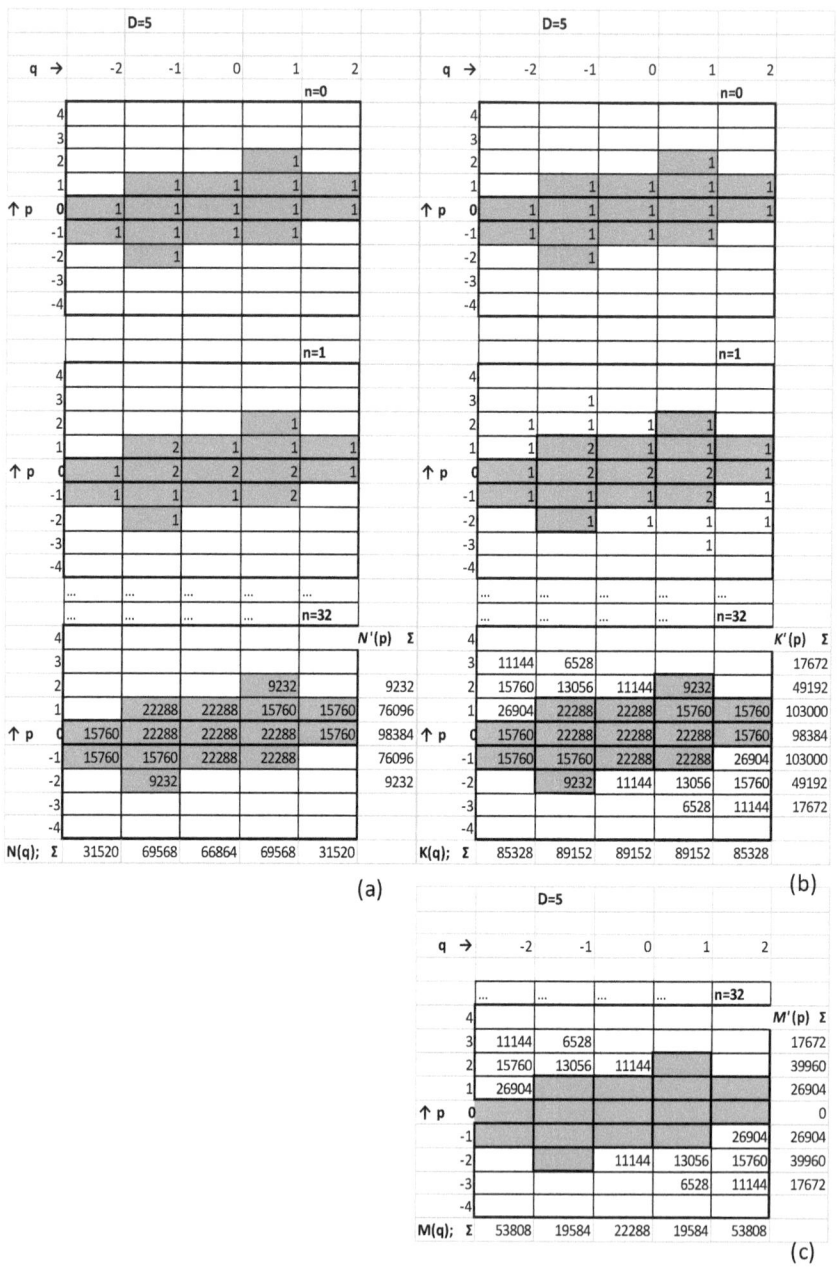

Fig.25. Calculation of filling with numbers of nonlinear arithmetic parallelepipeds of type 1.1 (a), 1.4 (b) and 1.5 (c) for a case $D = 5$ in the Excel program.

Envelopes schedules of calculations of distributions of number of rays in our ray system (Fig.25) of N(q), K(q), M(q) on a section and of $N'(p), K'(p), M'(p)$ on an angle are given in Fig.26 for a case $D = 5$, (Fig.6e, f, g) and $n = 32$. As show our calculations, the form of envelope practically doesn't change approximately after the 15th pass.

Form of envelopes in Fig.26a, b coincide with form of envelopes on Fig.23; form of envelopes in Fig.26c, d not completely coincide with form of envelopes in Fig.9, because of distinction of initial conditions.

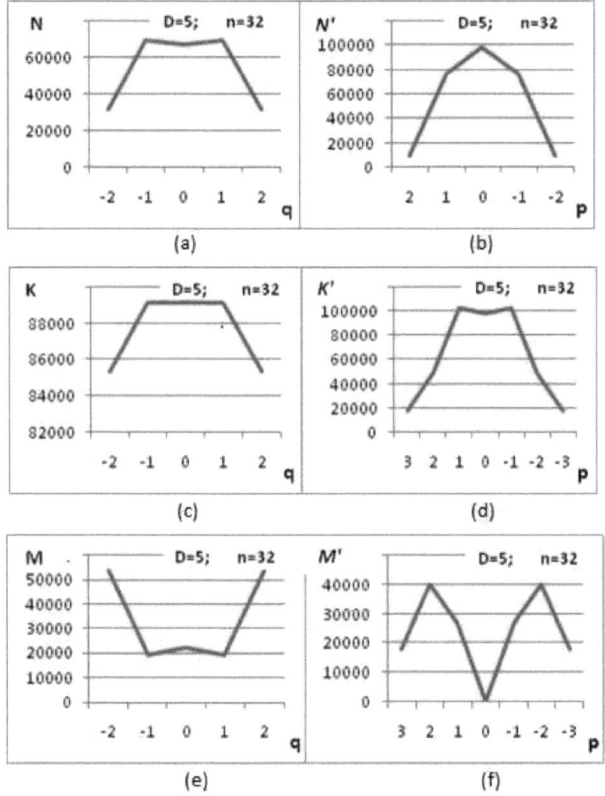

Fig.26 The envelopes of distributions of number of rays of K(q), N(q), M(q) on a section (a, c, e) and of $K'(p)$, $N'(p)$, $M'(p)$ on an angle (b, d, f) for a case of $D = 5, n = 32$.

Envelopes schedules of distributions of number of rays in our ray system (parallelepiped of type 1.1) of K(q) on a section and of K'(p) on angles are given in Fig.27 for a case $D = 255$ and $n = 0, 64, 256$. This case (Fig.27) is similar to the distribution represented on Fig.10 and Fig.11, the difference is in the set of initial conditions therefore the envelopes of distributions of rays at initial passes is differ. The form of envelope in Fig.27a, b, c is close to an ellipse (more precisely the top part of an ellipse), the form of envelope in Fig.27d is close to a parabola, the envelope in Fig.27e, f is close to Gaussian distribution.

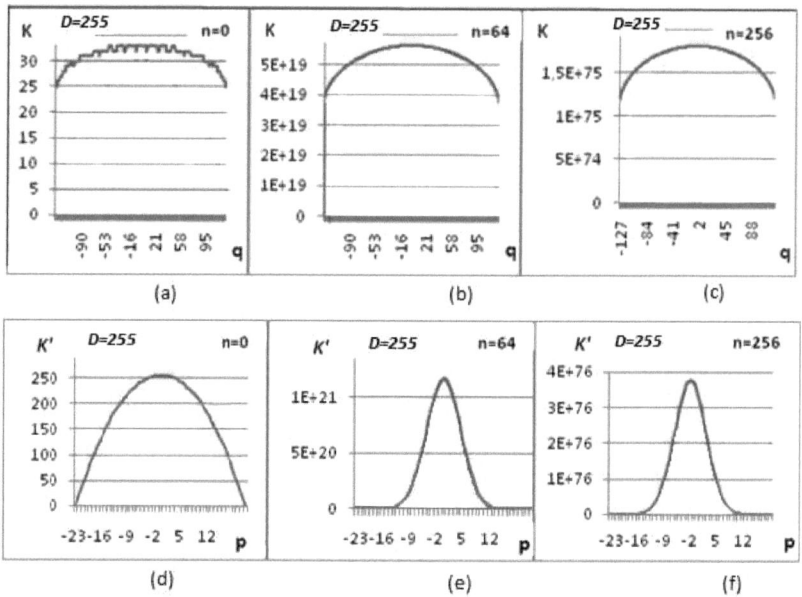

Fig.27. The envelopes of distributions of number of rays of K(q) on a section (a - c) and of K'(p) on an angle (d - f) for a case of $D = 255$.

Joint envelopes schedules of distributions of number of N(q), K(q), M(q) on a section (Fig.28) and of N'(p), K'(p), M'(p) on an angle (Fig.29) are given in Fig.28 and Fig.29.

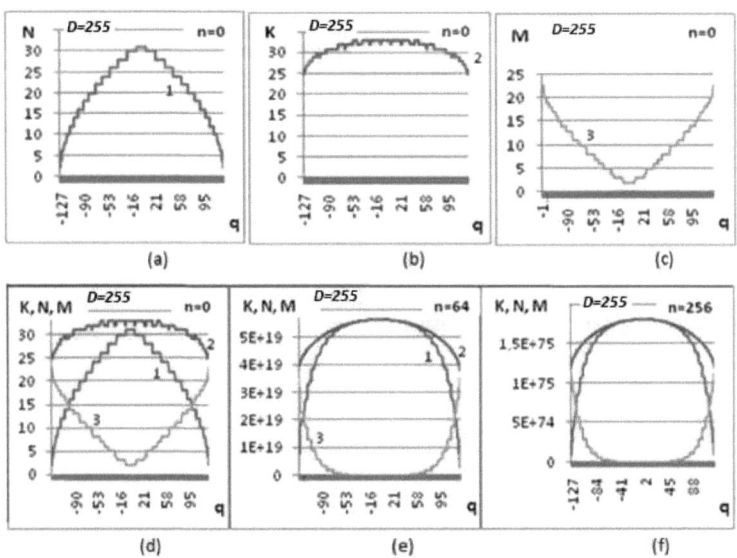

Fig.28. The envelopes of distributions of number of rays of N(q), K(q), M(q) on a section (a, b, c), curves 1, 2, 3, respectively, for $D = 255$, $n = 0$. Joint envelopes of distributions of number of rays of N(q), K(q), M(q) curves 1, 2, 3 (d, e, f), for $D = 255$, $n = 0, n = 64$, and $n = 256$.

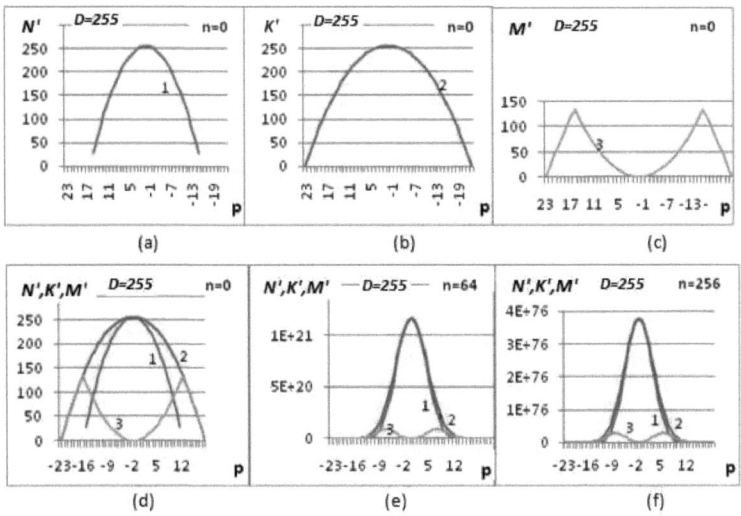

Fig.29. The envelopes of distributions of number of rays of $N'(p)$, $K'(p)$, $M'(p)$ on an angle (a, b, c), curves 1, 2, 3, respectively, for $D = 255$, $n = 0$. Joint envelopes of distributions of number of rays of $N'(p)$, $K'(p)$, $M'(p)$ curves 1, 2, 3 (d, e, f), for $D = 255$, $n = 0, n = 64$, and $n = 256$.

4. Angles and waves

As it was already noted in introduction, in work [5] wavy geometrical trajectories in the laser were described. Wavy trajectories consist of the links inclined on small angles γm, or γi, where m are natural numbers and i are integers.

In Fig.30 these "waves" are shown in the form of thick lines. Within length of the D of binary rays system one wave or packages of waves of different length can keep.

We will designate as λ_m the length of wavy trajectories; with increase of D λ_m grows discretely:

$$\lambda_m = 4Lm. \tag{41}$$

We will designate as v_m the height of this wave; the height v_m is proportional to the square of the length λ_m:

$$v_m \sim \lambda_m^2. \tag{42}$$

Numerical calculations for wavy trajectories showed that quantity of the rays extending along these waves or packages of waves are distributed unevenly on wavelength after a large number of n passes, i.e. at stationary distribution of rays in binary rays system. In Fig.30b we displayed this unevenness in the form of pieces of various "thickness" in various parts of a wave.

We assume that the energy of W extending along one link of trajectories of system of rays is proportional to number of rays of $\mathbb{N}$ imposed at each other along this link.

Energy of waves decreases with increase of wave length λ_m as energy (the number of rays is proportional to energy) is redistributed from longer waves to shorter waves.

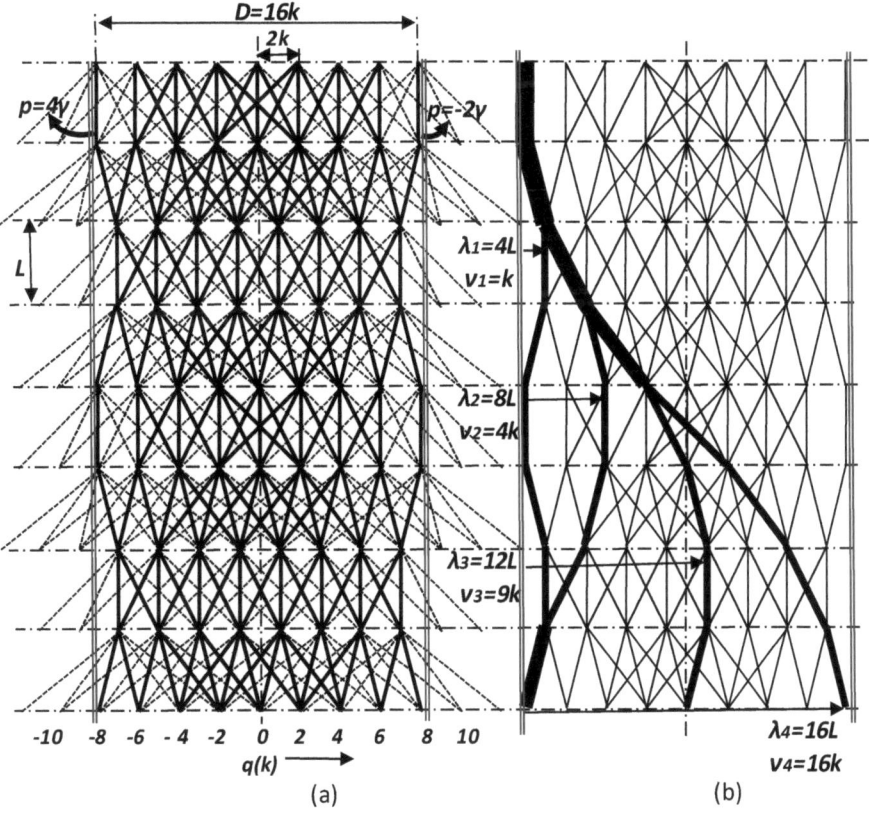

Fig.30. Binary rays system (a). Wavy trajectories (b). λ_m is length of "wave", v_m is height of "wave"; waves can be imposed at each other, places of imposing of waves are shown as increase in thickness of trajectories; various thickness of parts of a wave, corresponds to various number of the rays extending along various parts of a wave.

In Fig.31 packages of wavy trajectories for these "waves" and the angles characterizing these waves are shown to similar Fig.6f.

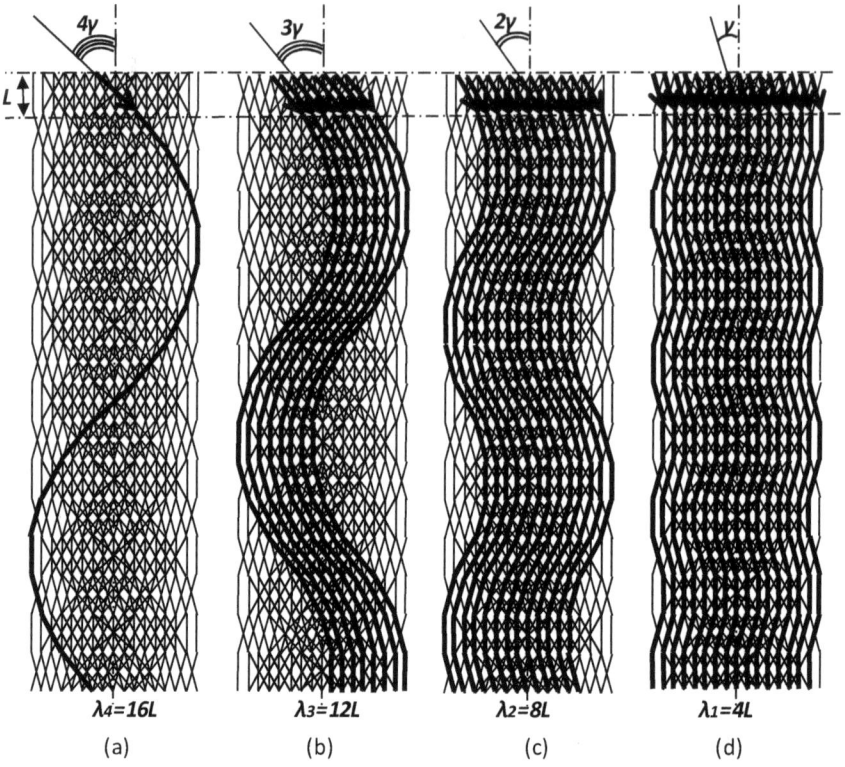

Fig.31. System (package) of wavy trajectories of various length $\lambda_m = 4Lm$ and angles γm or γi of inclination of links (and rays) of wavy trajectories are shown. Wave $\lambda_4 = 16L$ (a), system of waves $\lambda_3 = 12L$ (b), system of waves $\lambda_2 = 8L$ (c), system of waves $\lambda_1 = 4L$ (d).

We will give further the following reasonings. Our wavy trajectories consist of direct links. The shortest ($m = 1$) "wave" λ_1 (Fig.31) entering a package of such waves consists of 4 links inclined on angles 0 and γ. "Wave" ($m = 2$) λ_2 consists of 8 links inclined on angles $0, \gamma$ and 2γ. "Wave" ($m = 3$) λ_3 consists of 12 links inclined on angles $0, \gamma, 2\gamma$ and 3γ. "Wave" λ_4 consists of 16 links inclined on angles $0, \gamma, 2\gamma, 3\gamma$ and 4γ etc.

The longest "wave" λ_{max} it is λ_4 (or generally a package of such waves) differs from above-mentioned waves that only it incorporates the links inclined on the angle 4γ characterizing it. Feature of a wave or package of waves) λ_3 is that it incorporates the links inclined on the corner 3γ characterizing it and it is its difference from other, shorter waves etc.

From these reasonings it is logical to assume that distribution of rays on angle of $\mathbb{N}(p)$ (item 3.2.4) is proportional to distribution of energy of waves of $W(\lambda_m)$ depending on their length:

$$\mathbb{N}(p) \sim W(\lambda_m). \tag{43}$$

In Fig.31 it is visible that for this example the smallest energy has the longest wave $\lambda_{max} = \lambda_4 = 16L$ and the greatest has the system of the shortest waves $\lambda_1 = 4L$.

Thus:

$$\mathbb{N}(p) \sim W(\lambda_m) \sim 1/\lambda_m^2. \tag{44}$$

5. Half-integer rays system

In Fig.32 the other type of binary rays system is shown. The rays making this system are inclined on small angles:

$$p = (i + 1/2)\gamma, \tag{45}$$

where $i = 0, \pm 1, \pm 2 \ldots$ [4, 5]. We will call this group of rays as "$(i + 1/2)\gamma$ - system" or "half-integer (rays) system". For this binary rays system it is also possible to construct similar to the cases described in item 3 according to expressions (23, 25) another nonlinear arithmetic parallelepiped. Creation of such a parallelepiped is similarly to creation of a parallelepiped of $i\gamma$ – system (14).

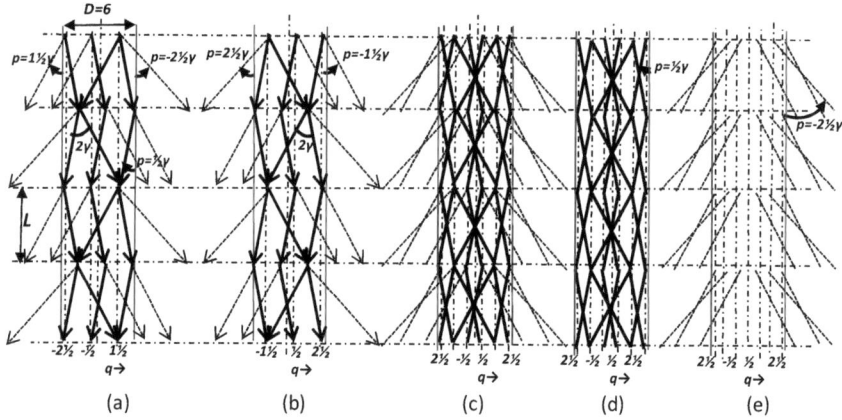

Fig.32. Binary ray $(i + 1/2)\gamma$ – system (half-integer system), height of nL and length of D; rays are inclined on angles $(i + 1/2)\gamma$. Construction begins with many units of a zero (top) row. (a, b) are 2 groups of the rays relating to $(i + 1/2)\gamma$ – system, (c) these 2 groups are combined together, (d) the periodic (wavy) trajectories which are a part of $(i + 1/2)\gamma$ – system represented on (c), (e) no periodic trajectories.

The binary rays system presented in Fig.32 contains periodic (wavy) and no periodic trajectories, length of wavy trajectories will be: $2L(2m + 1)$.

Integer $\{i\gamma$ – system$\}$ and half-integer $\{(i + 1/2)\gamma$ – system$\}$ the binary rays systems represented in Fig.6 and Fig.32, respectively, can serve additional evident geometrical interpretations of integer and half-integer respectively a spin of elementary particles [17].

6. Some general geometrical regularities and new opportunity of approximate evident interpretation of some physical processes

1. The form of the envelope of divergence of radiation of the single-mode laser in which light passes between mirrors infinite number of times [15] (Fig.33 of appendix) can be interpreted by means of the curve represented in Fig.9b, Fig.11e, f, Fig.14b, d, Fig.15b, d, Fig.27e, f, etc. (the form of the envelope is close to Gaussian distribution).

As a rule, the laser is supplied with the open laser resonator consisting of two plane-parallel mirrors. In this case, the small part of radiation leaves the bounds a laser aperture. (Compare Fig.33 with Figs.14, 15, border of an aperture is marked with dotted lines).

2. The form of the envelope of location of the particle in infinitely deep potential hole [16, 17] (Figs.34 - 36th of appendix) can be interpreted by means of the curve represented in Fig.14b, d, Fig.15b, d, Fig.24e, f, etc. (the form is close to Gaussian distribution). Borders of a ray system in Fig.14, Fig.15 and Fig.36 of apendix are marked with a dotted line.

The image of packages of waves (wavy trajectories) in a potential hole [16, 17] (Figs.34 - 36th of appendix) is represented in Fig.30 and Fig.31. The distribution of energy in waves of various lengths is proportional to a tilt angle of links. At zero pass of rays along links the distribution of rays is presented in Fig.24d and Fig.29a, d, etc. After many passes (iterations) the stationary distribution, in a form close to Gaussian distribution is reached, is presented in Fig.24e, f and Fig.29e, f, etc.

The form of envelope of angular distribution of links of trajectories is shown in Fig.24d and Fig.29a it is close to a parabola (i.e. in paraxial approach we obtain the square dependence) that corresponds to square dependence of distribution of power levels (Fig.34b of the appendix) and an electron wave form (Fig 35a of the appendix) [17].

The form of envelope of angular distribution of the rays extending along links of wavy trajectories in Fig.24e, f and Fig.29e, f is close to Gaussian distribution that corresponds to probability curve of finding of an electron in a potential hole [16, 17] (Fig.35b and Fig.36 of appendix).

The form of envelope of angular distribution of the rays in Fig.14 b, d and Fig.15b, d, is close to Gaussian distribution. A form of envelope of distribution of a small amount of rays which are going beyond a potential hole (beyond dashed lines in Figs.14, 15) can be interpreted as a geometrical model of tunnel effect [16, 17] (compare Figs.14, 15 with Fig.36 of the appendix). The form of envelope of distribution of this small amount of rays is close to an exponential curve [16, 17] (Fig.36 of appendix).

Perhaps various types (integer and half-integer) the described binary ray systems could be used for finding of various types of particles in a potential hole.

For example, on formal criterions the half-integer system represented in Fig.32 is more suitable for the description of finding of an electron in a potential hole, than the integer system represented in Fig.6 since the electron is characterized by half-integer spin [17].

3. The form of envelope of the liquid speed distribution on pipe section (Fig.37 of appendix) at a laminar [18] current can be interpreted by means of the curve represented in Fig.24a and Fig.28a. This curve in a form is close to a parabola [18] (Fig.37 of the appendix) or to parts of exponential curves of the second degree.

The form of envelope of the liquid volume distribution on pipe section at a laminar [18] current can be interpreted by means of the curve represented in Fig.24b, c and Fig.28 e, f. This curve in a form is close to a parabola of the fourth degree [18] or to parts of exponential curves of the fourth degree.

A laminar current can be interpreted with wavy trajectories in Fig.5b, Fig.6f, Fig.30b and Fig.31.

4. The form of envelope of the liquid speed (it is possible also volume) distribution on pipe section (Fig.38 of appendix) at a turbulent [18] current can be interpreted by means of the curve represented in Fig.10e, f, Fig.27a - c and Fig.28b, d, e, f. This curve in a form is close to an ellipse (more precisely to a half-ellipse, the top part of an ellipse) or "to a root of the seventh degree from distance to a wall" [18].

A turbulent current can be interpreted with joint consideration of wavy periodic and no periodic trajectories in Fig.6a – e and Fig.30a

7. Conclusion

By means of the above described models of binary rays system and a nonlinear arithmetic parallelepiped perhaps, very approximately and formally, however simply and geometrically visually to interpret above the given some physical processes.

8. References

1. *A. V. Yurkin.* New mirror for a laser resonator // *Sow. J. Quantum Electron.*, v. 21, p. 447, 1991.

2. *A. V. Yurkin.* Feasibility of reduction laser divergence // Sov. J. Quantum Electron., 1991, v. 21, p. 1096.]

3. *A. V. Yurkin.* Geometric features of a laser resonator consisting of many tilted reflecting planes //Sov. J. Quantum Electron., 1992, v. 22, p. 760.

4. *A. V. Yurkin.* Recurrence calculation of laser divergence and refractive analog of a multilobe mirror // Quantum Electron., 1993, v. 23, p. 323.

5. *A. V. Yurkin.* Quasi-resonator a new interpretation of scattering in lasers // *Quantum Electron.*, v. 24, p. 359, 1994.

6. *S. L. Popyrin, I. V Sokolov, A. V. Yurkin.* Three-dimensional geometrical analysis and the characteristics of laser generation in a multilobe mirror cavity // Optics Communications, 1999, v. 164, pp. 297 - 305.

7. *M. B. Mensky, A. V. Yurkin..* The `diffusion' of light and angular distribution in the laser equipped with a multilobe mirror // Procedings of Institute of Systems Analysis of RAS, 2008, v. 32, no. 2, pp. 113 – 121. arXiv:physics/0108037

8. *A. V. Yurkin.* System of rays in lasers and a new feasibility of light coherence control // Optics Communications, 1995, v. 114, p. 393.

9. *A. V. Yurkin.* The ray system in lasers, non-linear arithmetic pyramid and non-linear arithmetic triangles // Proceedings of the Institute of Systems Analysis of RAS, 2008, v. 32, no. 2, pp. 99 – 112 (Russian). arXiv:1302.5214

10. *A. V. Yurkin.* Ray trajectories and the algorithm to calculate the binomial coefficients of a new type // Proceedings of the Institute of Systems Analysis of RAS, 2009, v. 42, no.1, pp. 66 – 77 (Russian). arXiv:1302.4842

11. *A. V. Yurkin.* New view on diffraction discovered by Grimaldi and Gauss beams // Proceedings of the Institute of Systems Analysis of RAS, 2012, v. 62, no. 4, pp.28 – 35 (Russian). arXiv:1302.6287

12. *A. V. Yurkin.* New binomial and new view on light theory. About one new universal descriptive geometric model. (Lambert Academic Publishing, 2013). ISBN 978-3-659-38404-2.

13. *N. J. A. Sloane. S. Plouffe.* The Encyclopedia of Integer Sequences. New York: Academic Press, 1995. http://oeis.org/Seis.html. Sequences A053632

14. *A. N. Kolmogorov, I.G. Zhurbenko, A.V. Prokhorov.* Introduction to the theory of probability. Moscow: Nauka, 1995. (Russian).

15. *O. Svelto.* Principles of Lasers. Moscow: Mir, 1990. (Russian) [N.Y., 1989.]

16. *K. A. Putilov, V. A. Fabrikant.* Physics course. Moscow: Fizmatgiz, 1960, v. 3, ch. 9 (Russian).

17. *I. V. Savelyev.* Course of the general physics. Moscow: Nauka, 1982, v. 3, ch. 4 (Russian).

18. *K. A. Putilov.* Physics course. Moscow: Fizmatgiz, 1954, v. 1 (Russian).

Appendix

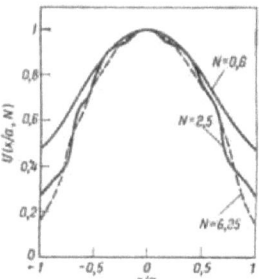

Fig.33. The mode amplitude of lower degree in plane-parallel laser resonator for three values of Fresnel number. An example of light field amplitude after about 200 passes of light between mirrors. Here N is the dimensionless number of Fresnel which is often applied in geometrical optics, $N = a^2/L\lambda$ where $2a$ is a laser aperture, L is distance between plane-parallel mirrors, and λ is wavelength. (Fig.4.21 from the monograph [15]).

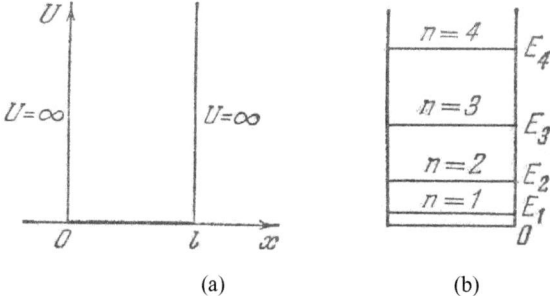

(a) (b)

Fig.34. The scheme of potential energy U of one-dimensional infinitely deep potential well (a); the scheme of the energy E_n levels in this well in the form of an "not equidistant energy levels", distance between which steps increase linearly as $(2n + 1)$, and the distance from a floor increases to steps quadratically in proportion to n^2 in the direction of from below up (b). (Fig.23.1 is from the monograph [17]).

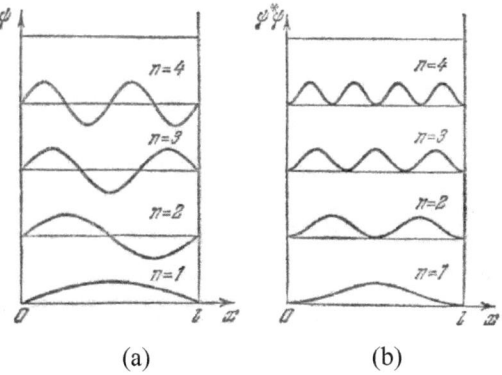

(a)   (b)

Fig.35. The scheme of the wave function ψ in one-dimensional infinitely deep potential well (a); and the scheme of density of probability of $\psi * \psi$ of detection of an electron at various distances from walls of this well (b), the scheme is provided in the form of an "equidistant energy levels ", distance between which steps constantly, and the distance from a floor increases to steps linearly in proportion to n. (Fig.23.2 is from the monograph [17]).

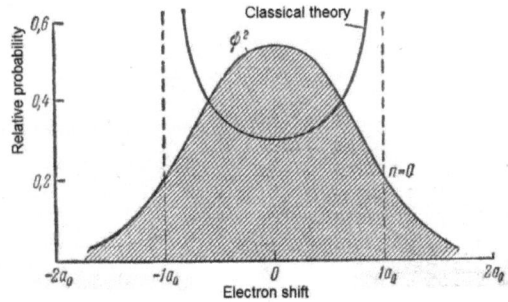

Fig.36. Relative probability to meet an electron at various distances from position of balance when it makes fluctuations, being in the lowest quantum state is shown. It is meant the classical theory of pendulum swing as the speed of swing of a pendulum is not equidistant, the probability of its detection near walls sharply increases (Fig.245 is from the monograph [12]).

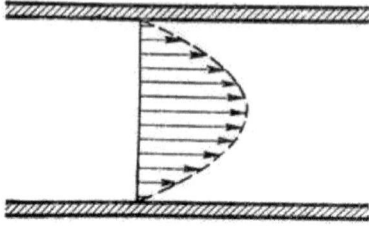

Fig.37. "... Speeds of a laminar current in a pipe are distributed under the parabolic law." [18, p. 284 - 285] "Therefore, under Puazel's law the amount of the liquid proceeding in 1 sec. on a pipe under the general equal conditions is proportional to the fourth degree of radius of a pipe." [18, p. 284 - 285]. (Fig.179 is from the monograph [18]).

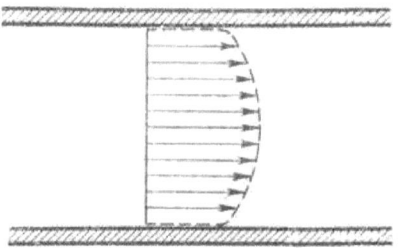

Fig.38. "At turbulent flow (Fig.180) current speed as shows experiment, is proportional approximately to a root of the seventh degree from distance to a wall:" [18, p. 284 - 285]. (Fig.180 is from the monograph [18]).

Printed by Books on Demand GmbH, Norderstedt / Germany